Plant-Microbe Interactions

Volume 2

Plant-Microbe Interactions
Volume 2

Edited by

Gary Stacey
University of Tennessee, Knoxville

Noel T. Keen
University of California, Riverside

SPRINGER-SCIENCE+BUSINESS MEDIA, B.V.

Cover design: Trudi Gershenov

Originally published by Chapman & Hall in 1997
Softcover reprint of the hardcover 1st edition 1997

1 2 3 4 5 6 7 8 9 10 XXX 01 00 99 98 97

Library of Congress Cataloging-in-Publication Data

Plant-microbe interactions / edited by Gary Stacey and Noel Keen.
p. cm.
Includes bibliographical references and index.
ISBN 978-1-4613-7772-6 ISBN 978-1-4615-6053-1 (eBook)
DOI 10.1007/978-1-4615-6053-1
1. Plant-microbe relationships. I. Stacey, Gary, 1951-
II. Keen, Noel, 1940-
QR351.P573 1995 95-10088
581'.2--dc20 CIP

British Library Cataloguing in Publication Data available

CONTENTS

Preface for Plant–Microbe Interactions Series

Anyone who reads *Science* and *Nature* regularly must realize that the field of plant–microbe interactions is hot. This is not totally unexpected since a close look at the complex stageplay involving plants and associated organisms was initiated only some twenty years ago. Because the performance has been playing for several million years, however, it is also not surprising that complex dramatic scenes involving parasitism, commensalism, and mutualism developed. We now occupy orchestra seats from which we can observe this drama, because the field of plant–microbe interactions was one of the first to deploy recombinant DNA approaches forcefully in the dissection of biology.

For example, the discovery of the elegant interkingdom transformation system in the crown gall disease rates as one of the great scientific revelations of this century. In the same sense, isolation of the nod signals in Rhizobium–plant interactions and the mechanisms for their synthesis comprise another amazing example of how microbes impose their will on plants. Since the stageplay of microbes and plants began eons ago, plants have not lain down and died. The identification of pathogen elicitors and the cloning and characterization of several plant disease resistance genes are providing crucial information on how disease resistance works and offer the chance to manipulate it.

Virulence mechanisms in pathogens, although studied for more than one hundred years, have become much better defined in the last ten years. Some mechanisms (e.g., certain pectic enzymes) that had been thought important for decades have now been convincingly proved to be so by the use of marker exchange mutant strains. Other putative virulence mechanisms, however, do not seem so important based on similar genetic tests, and in fact may be important in other pathogen life cycle stages. Our understanding of resistance mechanisms in plants has also benefited from the use of cloned genes and marker exchange mutants in pathogens. Such studies have recently provided the first solid evidence that preformed and inducibly formed secondary plant metabolites indeed are important defense agents, as has been postulated for many years.

It is rather amazing to savor historic contributions from the study of plant–microbe interactions to basic plant biology. Indeed, the long-postulated role of auxin/cytokinin ratios in determining plant morphogenesis received its most solid proof from studies of *Agrobacterium* strains mutated in particular T-DNA genes. Further, studies of plant growth hormones such as the cytokinins and gibberellins owe a great deal to the fact that certain microbial pathogens produce them copiously during disease development.

Other examples include the use of particular pathogen toxins and elicitors as metabolic probes and contributions of pathogens to studies of plant translocation patterns. Although the successes noted above are impressive, plant-microbe interactions is a young field and much excitement is yet to come. We are at the point, moreover, where these interactions can be manipulated for particular outcomes, including plant disease resistance and other plant improvements, environmental enrichment, new pharmaceuticals for humans and, of the greatest importance, more comprehensive understanding of basic plant biology. This book series is intended to chronicle these developments as they occur. Let the next act of the stageplay proceed. We are watching.

Preface

Plant–Microbe Interactions, Volume 2

Volume 1 of this series has made its appearance and dealt forcefully with important current topics in the field of plant–microbe interactions. We believe that the quality of those chapters was high and should serve as a focal point for the state of the art as well as an enduring reference. Volume 2 builds upon these accomplishments.

Chapter 1 discusses the fascinating lipo-chitin signal molecules from Rhizobium, aspects regarding their biosynthesis, and the basis for host specificity. These molecules are a cardinal example of how microorganisms influence plant development and stimulate speculation that they have identified a previously unknown aspect of plant hormone activity. Chapter 2 continues the discussion of Rhizobium by considering the trafficking of carbon and nitrogen in nodules. Although the ostensible advantage of nodules to plants is the fixation of atmospheric nitrogen, the actual process involved in supplying reduced nitrogen to the plant host is complex.

Chapter 3 addresses pathogen avirulence genes and available information on the biochemistry of avirulence gene activity in resistant plant hosts. The search for elicitors from some avirulence genes has been difficult but is certain to eventually unearth novel mechanisms of interorganismal signaling. Chapter 4 is concerned with the use of new approaches to studying a classic proposition in plant pathology, namely, the role of preformed toxic compounds in plant disease resistance. By skillfully employing molecular genetics with fungal pathogens, the authors and others have developed rather compelling evidence supporting the role of saponins as bona fide defense agents in plants.

Chapter 5 is concerned with the biology and epidemiology of a very important group of DNA plant viruses, the geminiviruses. These "twin DNA" viruses not only have great economic importance but are fueling interest in areas ranging from infection processes to plant resistance. Also in the virus arena, chapter 6 discusses viral virulence and avirulence genes. Because viruses have relatively compact genomes, plants appear to have keyed on almost all viral protein products as elicitors to initiate active defense mechanisms. Because these same protein products are required for viral replication, cell-to-cell movement, or transmission, their ancillary role as avirulence genes offers an outstanding opportunity to study viral–plant evolution.

Noel Keen
Gary Stacey

Contributors

Guido V. Bloemberg
Leiden University
Institute of Molecular Plant Sciences
Wassenaarseweg 64
2333 AL Leiden
The Netherlands

J. K. Brown
The University of Arizona, College of Agriculture
Department of Plant Sciences
Building 36
Tucson, AZ 85721

J. N. Culver
Center for Agricultural Biotechnology
University of Maryland Biotechnology Institute
College Park, MD 20742

M. J. Daniels
Sainsbury Laboratory
John Innes Centre, Colney Lane
Norwich, NR4 7UH
United Kingdom

Hiroshi Kouchi
National Institute of Agrobiological Resources
Kannondai 2-1-2
Tsukuba, Ibaraki 305
Japan

Jan E. Leach
Department of Plant Pathology
Throckmorton Plant Sciences Center
Kansas State University
Manhattan, KS 66506-5502

Ben J. J. Lugtenberg
Leiden University
Institute of Molecular Plant Sciences
Wassenaarseweg 64
2333 AL Leiden
The Netherlands

A. E. Osbourn
Sainsbury Laboratory
John Innes Centre, Colney Lane
Norwich, NR4 7UH
United Kingdom

Herman P. Spaink
Leiden University
Institute of Molecular Plant Sciences
Wassenaarseweg 64
2333 AL Leiden
The Netherlands

Shigeyuki Tajima
Laboratory of Plant Biochemistry
Department of Bioresource Science, Faculty of Agriculture
Kagawa University
Ikenobe 2366, Miki-cho, Kita-gun
Kagawa 761-07
Japan

Jane E. Thomas-Oates
Bijvoet Center for Biomolecular Research
Department of Mass Spectrometry
Utrecht University
Sorbonnelaan 16
3584 CA Utrecht
The Netherlands

Frank F. White
Department of Plant Pathology
Throckmorton Plant Sciences Center
Kansas State University
Manhattan, KS 66506-5502

J. P. Wubben
Sainsbury Laboratory
John Innes Centre, Colney Lane
Norwich, NR4 7UH
United Kingdom

CHAPTER

1

Biosynthesis and Host Specificity of Rhizobial Lipo-Chitin Oligosaccharide Signal Molecules

Guido V. Bloemberg,
Jane E. Thomas-Oates,
Ben J. J. Lugtenberg,
and Herman P. Spaink

INTRODUCTION

Rhizobia are able to elicit the formation of a new organ, the nodule, on the roots of leguminous plants. Nodulation is a host-specific process in that each rhizobial species or biovariety nodulates a restricted set of plants (for examples, see Table 1-1). Over the last 10 years it has become clear that reciprocal signaling between rhizobia and their host plants determines the host specificity of nodulation. Legumes secrete flavonoids or isoflavonoids, which are able to activate the rhizobial regulatory nodulation protein NodD in a host-specific way.[1] The activated NodD protein is able to induce the expression of the other bacterial nodulation (*nod* and *nol*) genes. It has been shown that many nodulation genes are involved in the biosynthesis or secretion of Nod metabolites (Table 1-2), which function as signals for the host plant. The basic structure of these signal molecules is a ß-1,4 linked *N*-acetylglucosamine oligosaccharide of which the nonreducing terminal sugar residue is *N*-acylated, so that they are described as lipo-chitin oligosaccharides (LCOs) (Fig. 1-1). Since the publication of Lerouge et al. (1990)[2] describing

Table 1-1. Host Specificity of Nodulation

Bacterial Species	Host Plants	References
R. meliloti	*Medicago, Melilotus, Trigonella*	119
R. l. bv. *viciae*	*Pisum, Vicia, Lathyrus, Lens*	119
R. l. bv. *trifolii*	*Trifolium*	119
R. etli	*Phaseolus*	56
R. loti	*Lotus*	119
R. tropici	broad host range, *e.g., Phaseolus, Leucaena*	120
R. fredii	broad host range, *e.g., Phaseolus, Glycine*	18
R. sp. GRH2	broad host range, *e.g., Phaseolus, Acacia*	121
R. sp. NGR234	broad host range, *over 75 plant* genera, incl. the nonlegume *Parasponia*	67
B. japonicum	*Glycine, Macroptilium*	122
B. elkanii	*Glycine, Macroptilium*	122
A. caulinodans	*Sesbania*	123

the first structure of a nodulation signal molecule produced by *R. meliloti,* many other LCOs have been structurally identified (Table 1-3). Many bacterial strain-specific variations have been found in the substituents on the chitin backbone, such as the presence of polyunsaturated acyl moieties and an *O*-acetyl moiety. An overview of LCO structures is shown in Table 1-3. For several of the bacterial strain-specific substituents a host specificity-determining function has been identified.

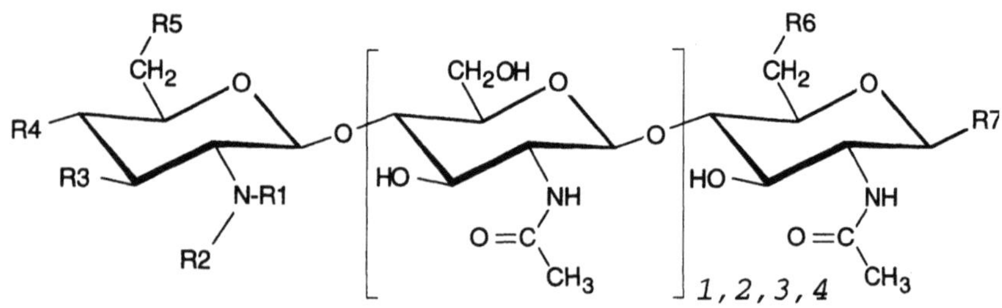

Figure 1-1. General structure of rhizobial lipo-chitin oligosaccharide (LCO) signal molecules. R1–R7 represent substituents that can vary in structure, of which an overview is given in Table 1-3.

Table 1-2. Biochemical Function of *nod* Gene Products Involved in Lipo-chitin Oligosaccharide Synthesis

nod Gene	Function	Evidence	References
nodM	glucosamine synthase	Sequence homology; complementation studies	20–22
nodA	*N*-acyltransferase	*In vitro* studies using crude cell extracts	13, 14
nodB	de-*N*-acetylase	Sequence homology; *in vitro* studies using purified protein; *in vivo* identification of intermediates	10–12
nodC	UDP-GlcNAc transferase	Sequence homology; *in vitro* studies using crude cell extracts; *in vivo* identification of intermediates	4–6,8, 9, 12
nodF	acyl carrier protein	Sequence homology; carries a 4'-phosphopantetheine prosthetic group; required for production of polyunsaturated fatty acyl moieties	32–34, 37
nodE	β-ketoacyl synthase	Sequence homology; involved in production of polyunsaturated fatty acyl moieties	3, 16, 17, 35, 37
nodL	*O*-acetyltransferase	Sequence homology; mutant studies; *in vitro* studies using purified protein	3, 41, 44, 45
nodP	ATP-sulphurylase subunit	Sequence homology; complementation studies; *in vitro* studies using crude cell extracts	49, 50
nodQ	ATP-sulphurylase subunit, APS kinase	Sequence homology; complementation studies; *in vitro* studies using crude cell extracts	49, 50
nodH	sulfotransferase	Sequence homology; mutant studies; *in vitro* studies using partially purified protein	14, 48, 51
nodZ	fucosyltransferase ?	Mutant studies	53
nodS	methyltransferase	Sequence homology; mutant studies; *in vitro* studies using partially purified NodS protein	52 Geelen, pers. comm.
nodX	*O*-acetyltransferase	Sequence homology; mutant studies	46, 47

Table 1-3. Structural Variations of Rhizobial Lipo-Chitin Oligosaccharides

	Substituents of Lipo-Chitin Oligosaccharides[a]							
Species	R1	R2	R3/R4	R5	R6	R7	n	References
R. meliloti	H	C16:1, C16:2, C16:3, (ω-1)OH-C18:0, (ω-1)OH-C20:0, (ω-1)OH-C22:0, (ω-1)OH-C24:0, (ω-1)OH-C26:0	H	Ac, H	S	H	1,2,3	2, 15, 31, 40
R. l. bv. *viciae*	H	C18:0, C18:1, C18:4 C16:0, C16:1	H	Ac	H, Ac	H	2,3	3, 16, 46
R. l. bv. *trifolii*	H	C18:0, OH-C18:0, C18:1, C18:3, C20:1, C20:2, C20:3, C20:4	H	Ac	H	H	1,2,3	16
R. etli	Me	C18:0, C18:1	Cb(*O*-4), H	H	AcFuc	H	3	73
R. loti	Me	C18:0, C18:1	Cb(*O*-4), H	H	4-*O*-AcFuc	H	3	68
R. tropici	Me	C18:1	H	H	S, H	H	3	124
R. fredii	H	C18:1	H	H	Fuc, 2-*O*-MeFuc	H	1,2,3	18
R. sp. GRH2	H, Me	C16:0, C16:1, C16:2, C18:0, C18:1, C18:2, C20:1	H	H	H, S	H	2,3,4	19
*R.*sp. NGR234	Me	C16:0, C18:1	Cb (1 or 2), H	H	2-*O*-MeFuc, 2-*O*-Me,3-*O*-SFuc, 2-*O*-Me,4-*O*-AcFuc	H	3	64
B. japonicum	H	C16:0, C16:1, C18:1	H	Ac, H	2-*O*-MeFuc	H	3	27, 65
B. elkanii	H, Me	C18:1	H, Cb	Ac, Cb, H	2-*O*-MeFuc, Fuc	H, Gro	2,3	27
A. caulinodans	Me	C18:0, C18:1	H	Cb, H	H, D-Ara	H	2,3	66

Abbreviations: Ac, acetyl; D-Ara, D-arabinose; C, carbon; Cb, carbamoyl; Fuc, fucose; Gro, glycerol; H, hydrogen; Me, methyl.
[a]Positions of R1-R7 are depicted in Figure 1-1.

GENERAL FEATURES OF LIPO-CHITIN OLIGOSACCHARIDE BIOSYNTHESIS

The *nodABC* genes, present in all rhizobial species, are the only *nod* genes strictly required for the synthesis of a basic LCO, devoid of all bacterial strain-specific substituents.[3] NodC shows sequence homology with chitin synthases and cellulose synthases and has therefore been suggested to be involved in the synthesis of the chitin backbone.[4–6] Surprisingly, NodC shares most similarity with the eukaryotic protein DG42 from *Xenopus laevis,* the function of which is not known. In vitro and in vivo studies using crude cell extracts and radio-labeled UDP-*N*-acetylglucosamine or glucosamine have indicated that NodC indeed functions as a UDP-*N*-acetylglucosamine transferase, producing compounds that are chitinase sensitive and that comigrate on thin layer chromatography plates with chitin fragments of up to five residues in length.[7,8] Very recently, *R. leguminosarum* bv. *viciae nodC*-dependent compounds from a NodC-overproducing *E.coli* strain were purified and, using mass spectrometry, were identified as tri-, tetra-, and pentameric forms of ß-1,4-linked *N*-acetylglucosamine.[9]

The NodB protein has sequence similarity with fungal chitin de-*N*-acetylases.[10] Biochemical proof for NodB functioning as a chitin deacetylase was provided by John et al.,[11] who showed that purified NodB protein is able, in vitro, to deacetylate the nonreducing *N*-acetyl glucosamine residue of chitin fragments having a length of two, three, or four carbohydrate residues. Additional evidence for the functions of the NodB and NodC proteins is provided by the work of Spaink et al.,[12] who structurally identified *nodB* and *nodC*-dependent metabolites from a *Rhizobium* strain as pentameric and tetrameric chitin derivatives having a free amino group on the nonreducing terminal residue. Since the production of such compounds requires the combined presence of the NodB and NodC proteins, they are referred to as NodBC-metabolites. Combining the in vitro deacetylating activity of NodB observed by John et al.[11] with the demonstration that NodB is required for LCO synthesis it seems obvious that for the transfer of the acyl moiety to the nonreducing terminal residue of the chitin backbone, a free amino group is required (Fig. 1-2).

The NodA protein does not show significant sequence homology with any known protein. On the basis of in vitro experiments using crude cell extracts and labeled NodBC-metabolites, the NodA protein is postulated to function as an acyltransferase in LCO synthesis[13,14] (Fig. 1-2).

Most of the characterized LCOs have a glycan backbone length of four or five residues (see Table 1-3). For some rhizobial strains only pentameric or tetrameric LCOs have been reported and in the cases when both types are produced, their ratio is dependent upon the species or biovariety.[3,15–17] Minor amounts of trimeric LCOs have only been reported for *R. meliloti*, *R. fredii*, and *R. leguminosarum* bv. *trifolii*[15–18] and hexameric LCOs only for *R.* sp. GRH2.[19] The NodA, NodB, and NodC proteins are good candidates for determination of the differences

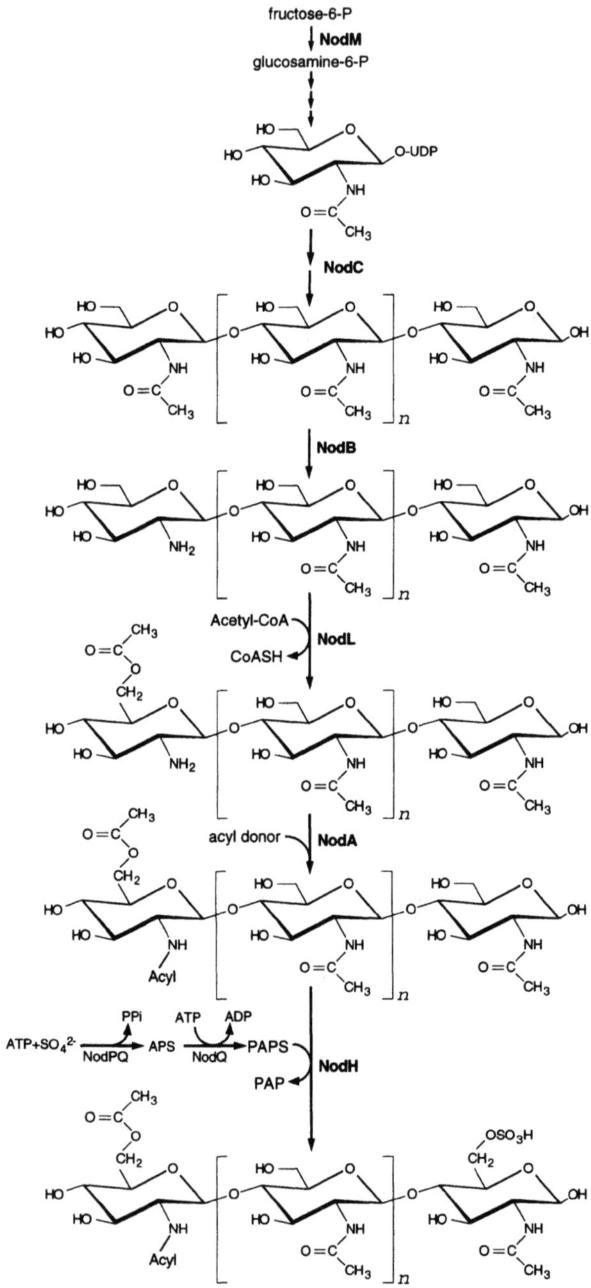
fructose-6-P
NodM
glucosamine-6-P
O-UDP
NodC
NodB
Acetyl-CoA
CoASH
NodL
acyl donor
NodA
Acyl
PPi
ATP
ADP
ATP+SO4 2-
NodPQ
APS
NodQ
PAPS
PAP
NodH
OSO3H

n=1,2,3 or 4

observed in glycan backbone lengths. Preliminary studies of NodA proteins from different rhizobial species indicate a role for the NodA protein in determining oligosaccharide backbone length[14] (Ritsema et al., unpublished).

UDP-*N*-acetylglucosamine is used by chitin synthases and by NodC as the substrate for polymerization. UDP-*N*-acetylglucosamine is synthesized in the bacterial cell as a precursor for lipidA and peptidoglycan, which explains why LCO synthesis does not require additional *nod* genes for the synthesis of UDP-*N*-acetylglucosamine. However, the *nodM* gene product shows sequence similarity with the *glsM* gene product, which functions as a glucosamine synthase.[20] Furthermore, *nodM* is capable of complementing *glsM*$^-$ mutant strains of *E. coli* and *R. leguminosarum*,[20–22] supporting this suggested function. The observation that in a *Rhizobium nodM* mutant strain quantitatively fewer LCOs are produced[23] supports the proposed function of NodM in increasing UDP-*N*-acetylglucosamine synthesis and thereby increasing LCO synthesis. There are no indications for the function of the *nodN* gene, which is located in the same operon as the *nodM* gene. However, on the basis of its genetic localization, *nodN* might have a function in UDP-*N*-acetylglucosamine synthesis.

By analogy with the known biosynthetic pathways for various oligosaccharides and polysaccharides, Carlson et al.[24] discussed extensively the possibility that, during LCO synthesis, the sugar oligosaccharide backbone is anchored in the cytoplasmic membrane (CM) by a glycosidically linked lipid carrier attached to the terminal carbohydrate residue. This carrier would be enzymatically cleaved off after completion of the biosynthesis of the LCO. The association of the NodC protein with the cytoplasmic membrane[25] supports such a hypothesis. Pyrophosphoryl undecaprenol, the lipid carrier in peptidoglycan biosynthesis,[26] has been suggested as a candidate. However, experimental data indicating the presence of LCOs linked to such a carrier are absent. For *B. elkanii* and a *nolO* mutant strain of *B. japonicum* a glycerol moiety glycosidically linked to the reducing *N*-acetylglucosamine has been found in a minority of the LCOs.[27,28] This glycerol moiety has been suggested to function as a carrier during synthesis or transport of the LCO.[24]

◀ **Figure 1-2.** Proposed model for the biosynthesis of lipo-chitin oligosaccharide signal molecules *in vivo* (for references, see Table 1-2). The following steps are depicted: (1) NodM functions as a glucosamine synthase in one of the early steps of UDP-*N*-acetyl glucosamine synthesis. (2) UDP-*N*-acetyl glucosamine is the substrate for NodC, which functions as a UDP-*N*-acetylglucosamine transferase in producing chitin fragments. (3) NodB de-*N*-acetylates the chitin fragments at the nonreducing terminal residue. (4) NodL transfers an acetyl group from acetyl-CoA to the C6-OH of the nonreducing terminal residue. (5) NodA transfers an acyl moiety from an acyl carrier (e.g., NodF or ACP) to the free amino group on the nonreducing terminal residue. (6) The NodP and NodQ proteins produce PAPS, which is used by NodH as sulfate donor in the sulfation of LCOs.

VARIATION IN THE ACYL MOIETY OF LCOs

On their nonreducing terminal glycan residue, LCOs carry an *N*-acyl moiety, the structure of which is variable. In most cases this is a fatty acyl substituent also commonly present in the phospholipids of the bacterial membranes, such as vaccenoyl (C18:1), stearyl (C18:0), palmitoyl (C16:1), or palmityl (C16:0) (Table 1-3). Minor amounts of LCOs from *R. meliloti* carry (ϖ-1)-hydroxylated fatty acyl moieties (C18:0 to C26:0).[29] These fatty acids are presumably present in the cell as precursors in the biosynthesis of 27-hydroxyoctacosanoic acid, a rhizobial lipopolysaccharide (LPS) component.[30] The synthesis of these LCOs is under the control of the *nodD3* and *syrM* regulatory genes,[31] but the regulated genes involved are unknown. Recently, a hydroxylated C18:0 acyl moiety was also identified in one of the minor LCOs produced by *R. leguminosarum* bv. *trifolii*.[16]

In the cases of *R. meliloti* and *R. leguminosarum* bvs. *viciae* and *trifolii*, LCOs are produced bearing specific α,ß-poly-unsaturated fatty acyl moieties, which have characteristic UV-absorption maxima dependent upon the number of double bonds *trans*-conjugated to the carbonyl group. The major LCOs produced by *R. meliloti* carry a C16:2 or C16:3 acyl moiety,[2,15] whereas *R. leguminosarum* bv. *viciae* produces LCOs with the more hydrophobic C18:4 substituent[3] (Fig.1-3). In contrast, in *R. leguminosarum* bv. *trifolii* the majority of LCOs produced contain a common fatty acyl moiety, while the remaining LCOs represent a complex mixture containing C18:3, C20:2, C20:3, and C20:4 fatty acyl moieties.[16] All of the latter polyunsaturated fatty acyl groups are more hydrophobic than those present on the LCOs of *R. meliloti* and *R. leguminosarum* bv. *viciae* (Fig.1-3). In addition to variations in length and the degree of unsaturation, polyunsaturated fatty acyl moieties of *R. leguminosarum* bv. *trifolii* also vary in the presence or absence of a *cis*-double bond.[16] This *cis*-double bond is present in all the polyunsaturated fatty acyl moieties identified from *R. meliloti* and *R. leguminosarum* bv. *viciae*[2,3,15] (Fig.1-3). Recently, the presence of LCOs bearing C16:2 and C18:2 moieties was also reported for the broad host-range *Rhizobium* strain GRH2.[19]

Both the *nodF* and *nodE* genes are required for the production of LCOs bearing polyunsaturated acyl moieties.[3,16,17,29,32,33] The NodF and NodE proteins show sequence homology with acyl carrier proteins (ACP) and ß-keto acyl synthases (KAS), respectively, both of which are enzymes involved in fatty acid biosynthesis.[34–37] Geiger et al.[38] showed that, like ACP, NodF protein carries a 4'-phosphopanthetheine prosthetic group, which supports the hypothesis that NodF protein functions as a specialized ACP in the biosynthesis of highly unsaturated fatty acids. We showed that for *R. leguminosarum* bvs. *viciae* and *trifolii* a central domain of the NodE proteins containing only 44 nonconserved amino acids, determines the structure of the acyl moiety.[17]

Since the *R. meliloti nodG* gene is located in the same operon as the *nodFE* genes and the NodG protein shows homology with dehydrogenases and

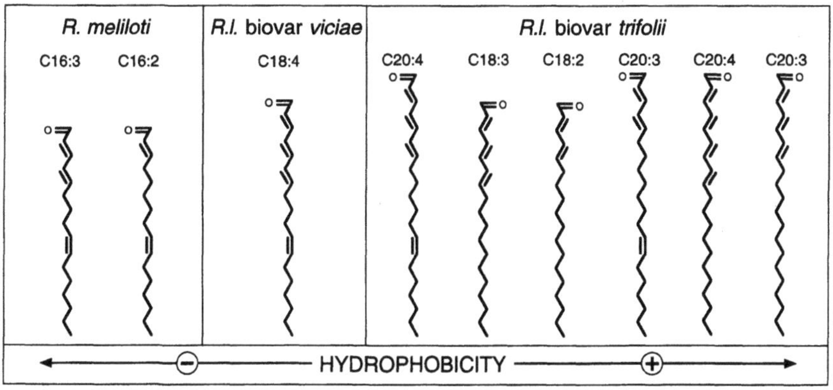

Figure 1-3. Chemical structures of polyunsaturated fatty acyl moieties of LCOs of *R. meliloti* and *R. leguminosarum* biovars *viciae* and *trifolii*, arranged according to their hydrophobicity.

ß-ketoacylreductases,[38,39] a function in the synthesis of the C16:2 and C16:3 fatty acyl chains was assumed. However, a mutation in *nodG* does not influence the production or the structure of the LCOs.[29]

ADDITIONAL BACTERIAL STRAIN-SPECIFIC DECORATIONS ON THE CHITIN BACKBONE

A variety of substituents of the nonreducing and reducing terminal sugar residue of LCOs has been identified. Substituents of hydroxyl groups include carbamoyl (R3, R4, R5), acetyl (R5, R6), sulfate (R6), fucose (R6), 2-*O*-methylfucose (R6), 2-*O*-Me,3-*O*-sulfatefucose (R6), 2-*O*-Me,4-*O*-acetylfucose (R6), 4-*O*-acetylfucose (R6), arabinose (R6), or glycerol (R7)(Fig.1-1; Table 1-3). Besides the characteristic acyl moiety (R2), a methyl group is the only other substitution on an amino group (R1)(Fig.1-1; Table 1-3).

The *nod* genes have been shown to be involved in the addition of several of the substituents listed above. The *O*-acetyl moiety, which is present on C6 of the nonreducing terminal residue of LCOs from *R. leguminosarum* bvs. *viciae* and *trifolii* and *R. meliloti,* is dependent on the presence of the *nodL* gene.[3,16,40] The NodL protein shows homology with acetyltransferases[41–43] and in vitro studies, using purified *R. leguminosarum* bv. *viciae* NodL protein, have shown that NodL functions as an *O*-acetyl transferase that transfers an acetyl group from acetyl-CoA to C6 of the nonreducing terminal residue of chitin fragments.[44,45] NodL protein can acetylate various substrates, such as LCOs, *N*-acetylglucosamine, chitin fragments, terminally deacetylated chitin derivatives, glucosamine, chitosan fragments, and cellopentaose.[44,45] Substrate specificity studies show that a free amino group at the nonreducing terminal residue is preferred by the NodL protein,[45] indicating that NodBC intermediates are the in vivo substrates for NodL protein (Fig. 1-2).

R. leguminosarum bv. *viciae* strain TOM produces an additional pentameric LCO bearing a second *O*-acetyl group at the C6 of the reducing terminal carbohydrate residue, the presence of which is dependent upon the *nodX* gene.[46] The NodX protein shows homology with some *O*-acetyl transferases, but does not show significant homology with the *nodL* gene product.[47]

The *nodH, nodP,* and *nodQ* genes are involved in the sulfation of the LCOs from *R. meliloti.*[48] In vitro studies using partially purified NodP and NodQ proteins show that they form a sulphate-activating complex for the synthesis of 3'-phosphoadenosine 5'-phosphosulfate (PAPS)[49,50] (Fig.1-2). NodP and NodQ together specify ATP-sulfurylase activity and, in addition, NodQ has adenosine 5' phosphosulfate (APS) kinase activity.[50] These functions are consistent with the homologies of *nodP* and *nodQ* with the *E.coli cysD, cysN,* and *cysC* genes.[49,50] For efficient sulfation, two copies of the *nodPQ* genes located on different megaplasmids in *R. meliloti* must be present.[48] The NodH protein shows homology with mammalian sulfotransferases and a mutation in the *nodH* gene prevents sulfation of LCOs.[48] Recently, in vitro NodH-dependent sulfotransferase activity was shown using partially purified NodH protein.[14,51] Kinetic analyses revealed that NodH has a higher affinity for LCOs than for chitin fragments or their terminally de-*N*-acetylated derivatives, indicating that LCOs are the in vivo substrates for NodH protein[51] (Fig. 1-2).

LCOs from various rhizobial strains are *N*-methylated on the nonreducing terminal sugar residue (Table 1-3). The *nodS* gene, identified in *R.* sp. NGR234, *R. etli, R. fredii, B. japonicum,* and *A. caulinodans,* shows similarity with a heterogenous group of S-adenosylmethionine (SAM)-utilizing methyltransferases.[52] For *A. caulinodans* it has been shown that the NodS protein can bind SAM and that a mutant *nodS* strain was not able to produce detectable amounts of methylated LCOs.[52] Recently, in vitro studies using partially purified NodS protein or a purified fusion NodS protein indicated that NodS is able to methylate NodBC metabolites, but not chitin fragments or nonmethylated LCOs (D. Geelen, pers. comm.). These results indicate that the NodS protein functions as a methyltransferase in LCO synthesis.

From mutant studies it was shown that the *nodZ* gene, identified in *R.* sp. NGR234 and *B. japonicum,* is involved in fucosylation of LCOs.[53] So far no sequence similarity of the NodZ protein with any other known protein has been reported.

SECRETION OF LIPO-CHITIN OLIGOSACCHARIDES

The molecular mechanism of LCO secretion is not known. However, it has been postulated that LCO secretion is mediated by the NodI, NodJ, and NodT proteins that show sequence homology with components of the ATP-dependent transport complexes of gram-negative bacteria, in particular those mediating the secretion of capsular polysaccharides.[54–57] NodI shows homology with ATP-binding cassette (ABC) proteins and NodJ with associated cytoplasmic membrane proteins. The association of the NodI protein with the cytoplasmic membrane[58] is consistent with the cellular localization of its homologues. The NodT protein sequence reveals a typical N-terminal signal peptide followed by a potential lipoprotein peptidase cleavage sequence and shows homology with a particular class of outermembrane protein components of ATP-dependent transport complexes.[57] Mutant studies of the *nodI* or *nodJ* genes in *R. leguminosarum* bv. *trifolii* showed that in both mutants LCOs were produced but were not exported out of the cell nor into the periplasmic space, indicating a role of these proteins in transport of LCOs.[59] For *R. leguminosarum* bv. *viciae* initial experiments suggested that the absence of the *nodIJ* genes, and also of the *nodT* gene, does not influence the secretion of LCOs.[23] However, studies show that the presence of the *nodIJ* genes specifically stimulates LCO secretion during the first few hours after *nod* gene induction in *R. leguminosarum* bv. *viciae* and *R. leguminosarum* bv. *trifolii*.[60] These results seem to be in contrast with the results of McKay and Djordjevic.[59] The presence of chromosomal counterparts of the *nodI, nodJ,* and *nodT* genes in *R. leguminosarum* bv. *viciae*[61] (A.H.M. Wijfjes, pers. comm.) could explain why *nodI* and *nodJ* were found not to be essential for the secretion of LCOs.

BIOLOGICAL EFFECTS OF LCOs

Exogenously applied LCOs can induce various discernable effects on plant roots, some of which can also be observed during the nodulation process. Applied in a concentration range of 10^{-8} to 10^{-12} M, LCOs can elicit various responses on root hairs in a relatively short time. *R. meliloti* LCOs applied to alfalfa roots induce, within 5 to 7 min., a depolarization of the root hair membrane, an increase in proton efflux, and an oscillating pattern of Ca^{2+} influx and efflux.[62,63] Furthermore, breakdown and rearrangements are observed in the actin filament network within 10–15 min.[63] Root hair deformation is a commonly observed effect of LCOs.[2,3,15,19,27,64–68] The only modification that has been shown to affect root hair deforming activity is the sulfate moiety on the reducing terminal residue of the LCOs of *R. meliloti*.[2] For *Vicia* plants, an incubation period of 5–10 min with LCOs is sufficient for induction of root hair deformation, which is visible after 1 h.[69] Only for *Macroptilium atropurpureum* it was shown that LCOs can evoke "shepherd's crook" curling of root hairs.[67] This curling effect requires LCO concentrations of 10^{-9} to 10^{-7} M and can also be induced by LCOs from rhizobia that do not nodulate *Macroptilium*, indicating that this response is not dependent upon host biovar-specific substituents.[67] The induction of responses corresponding to later stages in the nodule formation process require higher concentrations of

LCOs (at least 10^{-7} M) and are dependent upon the presence of host–strain-specific substituents. For example, formation of preinfection thread structures in *Vicia* plants induced by the LCOs of *R. leguminosarum* bv. *viciae* does require the presence of the *O*-acetyl group and the C18:4 polyunsaturated fatty acyl moiety.[70] On *Vicia, Medicago, Sesbania, Glycine, Trifolium, Lotus, Phaseolus,* and *Acacia* roots, host-specific LCOs in micromolar concentrations can trigger the division of cortical cells leading to the formation of a root nodule primordium.[3,17,19,66,68,71–73] For *Medicago, Sesbania,* and *Glycine* these primordia can even develop into full-grown nodules with an anatomical organization, similar to that observed in rhizobia-induced nodules.[66,71,72] The structure of the LCO-induced nodule primordium is indistinguishable from nodule primordia induced by bacteria. However, in some cases an abundant presence of severely deformed root hairs was observed on the LCO-induced nodule primordia, which distinguishes them from rhizobia-induced nodule primordial.[17,68] Furthermore, LCO-induced nodule primordia on *Lotus preslii* grow into a structure significantly different from a rhizobial-induced nodule structure.[68]

Several plant responses to exogenously applied LCOs, which are also observed at the molecular level during the rhizobial nodulation process, have been identified. For instance, early nodulin genes, such as *ENOD5, ENOD12,* and *ENOD40,* can be induced by LCOs in concentrations of 10^{-12} M, the spatial expression patterns of which are identical to the rhizobia-induced patterns.[74–78] *ENOD12* induction was shown to occur 3 h after LCO application.[79] When *Vicia* roots are unshielded from light, *R. leguminosarum* bv. *viciae* can increase flavonoid synthesis and in addition induce the production of new flavonoids.[80] These changes are concomitant with the induction of phenylalanine lyase (PAL) and chalcone synthase (CHS), which are both key enzymes in flavonoid synthesis.[81–82] The increased level of flavonoid synthesis can be measured by an increase in the *nod* gene-inducing activity (INI) of root exudates.[83] Host-specific *R. leguminosarum* bv. *viciae* LCOs in concentrations of 10^{-7} M are also able to induce such an INI response on *Vicia* plants.[3] Recently, elevated levels of flavonoids were also observed in root exudates of *Glycine* after application of LCOs from *Bradyrhizobium japonicum* or *R.* sp. NGR234.[84]

A curious effect shown for LCOs from *R. leguminosarum* bv. *viciae* is that they can block nodulation on *Vicia* by rhizobia when applied before the addition of bacteria or when applied in large quantities.[85] These effects, which are referred to as the *JAN-phenotype* (jamming of nodulation), indicate that LCOs may play a role in autoregulation of nodulation. Relić et al.[86] have shown that LCOs from *R.* sp. NGR234, when applied in exogenous concentrations between 10^{-7} and 10^{-6} M permit nodulation on *Vigna* and *Glycine* of various *nodABC*$^{-}$ mutants of *R.* sp. NGR234, *B. japonicum,* and *R. fredii.* In the presence of LCOs these strains enter the root via normal infection threads in root hairs, which indicate that LCOs determine infection.

Host-specific LCOs of *R. meliloti* were also shown to activate the cell cycle of *Medicago* cells in suspension cultures.[87] Even at picomolar concentrations, LCOs significantly enhance the expression of cell cycle markers, thymidine incorporation, kinase activity of $p34^{cdc2}$-related complexes, and the number of S-phase cells.[87] At micromolar concentrations, LCOs also induce the expression of the isoflavone reductase gene (*IRF*), a marker gene of the isoflavonoid synthesis pathway.[87] However, this response can also be induced by exogenously applied chitintetraose, even in a concentration of 10^{-9} M.[87] The latter observation is consistent with the observation that chitin fragments can elicit defense responses, e.g., phytoalexin synthesis, in rice cell cultures.[88] LCOs from different rhizobia added to tomato cell culture suspensions at final concentrations of 10^{-8} M can stimulate a rapid transient alkalinization of the medium, showing that tomato, a nonhost plant for rhizobia, can perceive LCOs.[89] However, this response is also induced by chitin fragments at lower concentrations,[90] indicating that this perception system is characteristic for chitin fragments or their derivatives rather than for LCOs.

ROLE OF BACTERIAL STRAIN-SPECIFIC SUBSTITUENTS OF LCOs IN BIOLOGICAL ACTIVITY AND HOST SPECIFICITY

In several cases it has been shown that the host-specific character of LCOs is determined by bacterial strain-specific substituents. In the case of *R. meliloti* the NodH and NodPQ-determined sulfate group is of crucial importance for some of the biological activities of LCOs and for the determination of host range.[40,71] A lack of the sulfate group in the LCOs from *R. meliloti* NodH leads to a loss of nodulation on *Medicago,* but instead, these strains nodulate the nonhost *Vicia.*[40,91–93] *R. meliloti* mutant strains, which produce a mixture of sulfated and nonsulfated LCOs can nodulate both *Medicago* and *Vicia.*[40,94,95] The presence of the NodFE-determined polyunsaturated fatty acyl moiety and the NodL-determined *O*-acetyl group on the LCOs of *R. meliloti* seem to play a role in determining efficiency of nodulation.[96] The presence of at least one of these substituents in addition to the sulfate group is required for nodulation.[96] For *R. leguminosarum* bvs. *viciae* and *trifolii* it was shown that the difference in hydrophobicity of the NodFE-dependent polyunsaturated fatty acyl moieties determines the difference in host range between these biovars.[2,16,17] In both cases these polyunsaturated fatty acyl moieties are required for the induction of nodule primordia by purified LCOs on the host plants.[2,17] In addition, it was shown for LCOs produced by *R. leguminosarum* bv. *viciae* that the combined presence of the NodL-determined *O*-acetyl moiety on the nonreducing terminal residue and the C18:4 acyl moiety is required for the induction of nodule primordia and for the formation of preinfection threads on *Vicia* plants.[2,70] The *nodO* gene of *R. leguminosarum* bv. *viciae* can replace the *nodFEL* genes for nodulation functions.[20] The NodO protein, which is secreted by the bacterium,[97] has been shown to form cation-selective pores in membranes.[98] The molecular basis of the complementing ability of NodO has yet to be elucidated. Exogenously applied chitin oligosaccharides or their *O*-acetylated forms are unable to elicit any response similar to a stage of nodule formation on the host plant *Vicia.*[3,99] However, *O*-acetylated chitin tetrameric, pentameric, and hexameric oligosaccharides can induce nodule primordia on *Vicia* roots when applied by microballistic targeting.[99,100] The absence of the *O*-acetyl group decreases this effect drastically. From these results it can be hypothesized that the specific fatty acyl moiety has a role in targeting LCOs to the inner root tissue. The following observations support such a hypothesis: (1) polyunsaturated fatty acyl-containing LCOs seem only to play a role in the induction of indeterminate nodules, which are initiated in the inner cortex at a distance of several cell layers from the root surface and (2) the hydrophobicity of polyunsaturated fatty acyl moieties can be a major determinant of host specificity.[16,17]

Most strains of *R. leguminosarum* bv. *viciae* cannot nodulate *Pisum sativum* cv. Afghanistan.[101] However, *R. leguminosarum* bv. *viciae* strain TOM can nodulate Afghanistan pea, which nodulation is mediated by the presence of the additional *nodX* gene.[102] Recently, it was shown that NodX determines the presence of an additional *O*-acetyl moiety on C6 of the reducing terminal residues of LCOs, indicating that this additional substituent is able to extend the host range.[46]

The *nodS* gene of *Rhizobium* sp. NGR234 has been suggested to influence the efficiency of nodulation on *Leucaena* plants.[103] In addition, Krishnan et al.[104] have shown that the *nodS* gene of *Rhizobium* sp. NGR234 is capable of extending the host range of *Rhizobium fredii* strain USDA257 to include *Leucaena* spp. These results suggest that the NodS-determined *N*-methyl group[52] is involved in determining host specificity.

NodZ mutants of *B. japonicum* produce LCOs devoid of the fucose moiety and are defective in nodulating siratro plants, but not in nodulating *Glycine,* indicating a role for the fucose moiety in host range determination.[53] LCOs of *B. japonicum* bearing a 2-*O*-methylfucose or fucose can induce cortical cell divisions on *Glycine* roots. In contrast, LCOs devoid of such a fucose are no longer able to induce mitogenic activity, indicating that the presence of a fucose moiety has biological relevance for *Glycine.*[53] Recent results show that the presence of a fucose is necessary for biological activity of LCOs with a pentameric sugar chain, whereas tetrameric LCOs are only biologically active on *Glycine* when they are devoid of a substituent at the reducing terminus (K. Peters, pers. comm.). *B. japonicum* might also produce minor quantities of tetrameric LCOs, which could explain the contradictory results of Stacey et al.[53]

The biological relevance of other host biovar-specific substituents, such as D-arabinose or carbamoyl groups, has not yet been shown, but it is likely that they play a role in determination of the host range.

CONCLUDING REMARKS AND FUTURE PROSPECTS

Over the last 10 years many aspects of the molecular signaling between rhizobia and leguminous plants have been elucidated. A wide variety of rhizobial LCOs that have been shown to be major determinants of the host specificity of nodulation have been identified. For many *nod* and *nol* genes a function in synthesis or secretion of LCOs has been indicated. However, many aspects still have to be unraveled. Many genes involved in LCO synthesis still have to be identified, for example the genes involved in the substitution of carbamoyl, glycerol, and D-arabinose moieties. Very few data are available on the localization of LCO synthesis in the cell and the secretion mechanism of LCOs. A few Nod proteins have been purified and used for in vitro studies to show their biochemical function. At present detailed enzymatic studies have been performed only of the NodL protein[44,45] and the NodH protein.[54] Similar studies of the remaining Nod proteins are necessary to determine their precise functions in the biosynthesis of LCOs.

Almost nothing is known about the mechanism of perception of LCOs by plants. It has been shown that LCOs, when exogenously applied to *Vicia* roots, are rapidly degraded, probably by chitinases.[69,105] The presence of a NodH-determined sulfate group or a NodL-determined *O*-acetyl moiety has been shown to protect the LCOs of *R. meliloti* against breakdown by chitinases isolated from *Medicago* and *Vicia*.[105] Ardourel et al.[96] extensively discussed the possibility of the presence of multiple LCO receptors on root hairs. Observations that are supportive for a receptor model are (1) the host-range determining moieties on LCOs and (2) the ability of LCOs to induce effects on plants in extremely low concentrations. The recent characterization of a LCO binding site (NFBS1) in *Medicago* root preparations[106] is giving additional support to such a receptor hypothesis. However, the affinity of this site for LCOs is low in comparison with the concentrations of LCOs that can elicit responses on the host plant and it lacks a specificity for bacterial strain-specific LCO modifications.[106] Furthermore, an analogous LCO binding site was also identified in tomato root extracts.[106] To date, it is not clear if the identified binding site in *Medicago* has a role in nodulation. Previously, high affinity binding sites for chitin fragments have been identified in intact rice and tomato plant cells.[107,108] It was also shown that LCOs bind to these putative receptors,[108] which could explain the similar alkalanizing effect of chitin fragments and LCOs on tomato cell cultures.[89] The LCO-binding sites from *Medicago* and tomato characterized by Bono et al.,[104] are different from the chitin-binding site in tomato identified by Baureithel et al.,[108] since in the former case the sites have a higher affinity for LCOs and chitintetraose appeared to be a poor competitor. From the observation that *O*-acetylated chitin fragments are able to induce nodule primordia when applied by microballistic targeting into root cortical cells, it can be hypothesized that a specific acyl moiety is involved in the transport of the active part of the molecule inside the plant.[99,100] It has been hypothesized that the overall hydrophobicity of the fatty acyl moiety of LCOs is a major determinant of host specificity,[16,17] which supports the latter hypothesis. Microballistic targeting of signal molecules is likely to be of great importance in the study of LCO perception and the mechanism of root nodule formation.

LCOs represent a novel, and so far unique, class of signal molecules involved in nodule organogenesis. It has been hypothesized that LCOs mimic endogenous plant signal molecules involved in developmental processes. Indications that form the basis for this hypothesis are: (1) transgenic *Nicotiana* plants containing the rhizobial *nodA* and *NodB* genes are severely altered in growth and development,[109] (2) LCOs are able to rescue a *Daucus* somatic embryo mutant that can also be rescued by the external application of a 32kD endochitinase from wild-type embryos,[110,111] (3) the spontaneous formation of nodules on *Medicago* in the absence of rhizobia,[112] and (4) the correlation between the expression of plant chitinases and developmental processes.[113] Benhamou and Asseling[114] and Spaink et al.[115,116] have published results that indicate that lipophilic chitin derivatives are present in plants. Recently Wagner et al.[117] reported the presence of chitinous tissue in the fish *Paralipophrys,* which is the first example of the existence of chitin derivatives in vertebrates. Furthermore, it has been shown that the *Xenopus* DG42 protein shows significant homology with chitin synthases (especially to NodC) and is specifically expressed during embryogenesis.[5,118] To date, no chitin-containing molecules from any of these organisms have been structurally identified. However, based on these results, it is tempting to speculate that chitin derivatives also play a role in embryogenesis of vertebrates. Our knowledge of the biosynthesis of rhizobial LCOs will be a powerful tool in the search for novel chitinlike signal molecules in plants and animals.

ACKNOWLEDGMENTS

G.V.B. and J.E.T.-O were supported by The Netherlands Foundation for Chemical Research (with financial aid from the Netherlands Organization for Scientific Reseach). H.P.S. was supported by the Royal Netherlands Academy for Arts and Sciences and a Pioneer Grant.

REFERENCES

1. Schlaman, H. R. M., R. J. H. Okker, and B. J. J. Lugtenberg. 1992. Regulation of nodulation gene expression by NodD in rhizobia. *J. Bacteriol.* **174:**5177–5182.
2. Lerouge, P., P. Roche, C. Faucher, F. Maillet, G. Truchet, J.-C. Promé, and J. Dénarié. 1990. Symbiotic host-specificity of *Rhizobium meliloti* is determined by a sulphated and acylated glucosamine oligosaccharide signal. *Nature* **344:**781–784.

3. Spaink, H. P., D. M. Sheeley, A. A. N. Van Brussel, J. Glushka, W. S. York, T. Tak, O. Geiger, E. P. Kennedy, V. N. Reinhold, and B. J. J. Lugtenberg. 1991. A novel highly unsaturated fatty acid moiety of lipo-oligosaccharide signals determines host specificity of *Rhizobium*. *Nature* **354:**125–131.
4. Atkinson, E. M. and S. R. Long. 1992. Homology of *Rhizobium meliloti* NodC to polysaccharide polymerizing enzymes. *Mol. Plant-Microbe Interact.* **5:**439–442.
5. Bulawa, C. E. and W. Wasco. 1991. Chitin and nodulation. *Nature* **353:**710.
6. Debellé, F., C. Rosenberg, and J. Dénarié. 1992. The *Rhizobium, Bradyrhizobium* and *Azorhizobium* NodC proteins are homologous to yeast chitin synthases. *Mol. Plant-Microbe Interact.* **3:**317–326.
7. Spaink, H. P., A. H. M. Wijfjes, O. Geiger, G. V. Bloemberg, T. Ritsema, and B. J. J. Lugtenberg. 1993. The function of the rhizobial *nodABC* and *nodFEL* operons in the biosynthesis of lipo-oligosaccharides. In *New Horizons in Nitrogen Fixation,* eds. R. Palacios, J. Mora, and W. E. Newton, pp. 165–170. Kluwer Academic Publishers, Dordrecht.
8. Geremia, R. A., P. Mergaert, D. Geelen, M. Van Montagu, and M. Holsters. 1994. The NodC protein of *Azorhizobium caulinodans* is an *N*-acetylglucosaminyltransferase. *Proc. Natl. Acad. Sci. USA* **91:**2669–2673.
9. Kamst, E., K. M. G. M. van der Drift, J. E. Thomas-Oates, B. J. J. Lugtenberg and H. P. Spaink. Mass spectrometric analysis of chitin oligosaccharides produced by *Rhizobium* NodC protein in *Escherichia coli.* J Bacterial. **177:**6285–6285.
10. Kafetzopoulos, D., G. Thireos, J. N. Vournakis, and V. Bouriotis. 1993. The primary structure of a fungal chitin deacetylase reveals the function for two bacterial gene products. *Proc. Natl. Acad. Sci. USA* **90:**8005–8008.
11. John, M., H. Röhrig, J. Schmidt, U. Wieneke, and J. Schell. 1993. *Rhizobium* NodB protein involved in nodulation signal synthesis is a chitooligosaccharide deacetylase. *Proc. Natl. Acad. Sci. USA* **90:**625–629.
12. Spaink, H. P., A. H. M. Wijfjes, K. M. G. M. Van der Drift, J. Haverkamp, J. E. Thomas-Oates, and B. J. J. Lugtenberg. 1994. Structural identification of metabolites produced by the NodB and NodC proteins of *Rhizobium leguminosarum. Mol. Microbiol.* **13:**821–831.
13. Röhrig, H., J. Schmidt, U. Wieneke, E. Kondorosi, I. Barlier, J. Schell, and M. John. 1994. Biosynthesis of lipooligosaccharide nodulation factors—*Rhizobium* NodA protein is involved in *N*-acylation of the chitooligosaccharide backbone. *Proc. Natl. Acad. Sci. USA* **91:**3122–3126.
14. Atkinson, E. M., M. M. Palcic, O. Hindsgaul, and S. R. Long. 1994. Biosynthesis of *Rhizobium meliloti* lipooligosaccharide Nod factors: NodA is required for an *N*-acetyltransferase activity. *Proc. Natl. Acad. Sci. USA* **91:**8418–8422.
15. Schultze, M., B. Quiclet-Sire, E. Kondorosi, H. Virelizier, J. N. Glushka, G. Endre, S. D. Géro, and A. Kondorosi. 1992. *Rhizobium meliloti* produces a family of sulfated lipo-oligosaccharides exhibiting different degrees of plant host specificity. *Proc. Natl. Acad. Sci. USA* **89:**192–196.
16. Spaink, H. P., G. V. Bloemberg, A. A. N. van Brussel, B. J. J. Lugtenberg, K. M. G. M. van der Drift, J. Haverkamp, and J. E. Thomas-Oates. 1995. Host specificity of

Rhizobium leguminosarum is determined by the hydrophobicity of highly unsaturated fatty acyl moieties of the nodulation factors. *Mol. Plant-Microbe Inter.* **8:**155–164.

17. Bloemberg, G. V., E. Kamst, M. Harteveld, K. M. G. M. van der Drift, J. Haverkamp, J. E. Thomas-Oates, B. J. J. Lugtenberg, and H. P. Spaink. 1995. A central domain of *Rhizobium* NodE protein mediates host specificity by determining the hydrophobicity of fatty acyl moieties of nodulation factors. *Mol. Microbiol* **16:**1123–1136.
18. Bec-Ferté, M.-P., H. B. Krishnan, D. Promé, A. Savagnac, S. G. Pueppke, and J.-C. Promé. 1994. Structures of nodulation factors from the nitrogen-fixing soybean symbiont *Rhizobium fredii* USDA257. *Biochemistry* **33:**11782–11788.
19. López-Lara, I. M., K. M. G. M. Van der Drift, A. A. N. Van Brussel, J. Haverkamp, B. J. J. Lugtenberg, J. E. Thomas-Oates, and H. P. Spaink. 1995 Induction of nodule primordia on *Phaseolus* and *Acacia* by lipo-chitin oligosaccharide nodulation signals from broad host range *Rhizobium* strain GRH2, Plant Mol. Biol. **29:**465–477.
20. Downie, J. A. and B. P. Surin. 1990. Either of two *nod* gene loci can complement the nodulation defect of a *nod* deletion mutant of *Rhizobium leguminosarum* bv. *viciae. Mol. Gen. Genet.* **190:**359–365.
21. Baev, N., G. Endre, G. Petrovics, Z. Banfalvi, and A. Kondorosi. 1991. Six nodulation genes of *nod* box locus 4 in *Rhizobium meliloti* are involved in nodulation signal production: *nodM* codes for D-glucosamine synthetase. *Mol. Gen. Genet.* **228:**113–124.
22. Marie, C., M.-A. Barny, and J. A. Downie. 1992. *Rhizobium leguminosarum* has two glucosamine synthases, GlmS and NodM, required for nodulation and development of nitrogen-fixing nodules. *Mol. Microbiol.* **6:**843–851.
23. Spaink, H. P., A. Aarts, G. Stacey, G. V. Bloemberg, B. J. J. Lugtenberg, and E. P. Kennedy. 1992. Detection and separation of Rhizobium and Bradyrhizobium Nod metabolites using thin-layer chromatography. *Mol. Plant-Microbe Interact.* **5:**72–80.
24. Carlson, R. W., N. P. J. Price, and G. Stacey. 1994. The biosynthesis of rhizobial lipo-oligosaccharide nodulation signal molecules. *Mol. Plant-Microbe Interact.* **7:**684–695.
25. Barny, M. A. and J. A. Downie. 1993. Identification of the NodC Protein in the inner but not the outer membrane of *Rhizobium Leguminosarum. Mol. Plant-Microbe Int.* **6:**669–672.
26. Whitfield, C. and M. A. Valvano. 1993. Biosynthesis and expression of cell-surface polysaccharides in gram-negative bacteria. *Adv. Microb. Physiol.* **35:**136–246.
27. Carlson, R. W., J. Sanjuan, R. Bhat, J. Glushka, H. P. Spaink, A. H. M. Wijfjes, A. N. N. Van Brussel, T. J. W. Stokkermans, K. Peters, and G. Stacey. 1993. The structures and biological activities of the lipo-oligosaccharide nodulation signals produced by Type I and Type II strains of *Bradyrhizobium japonicum. J. Biol. Chem.* **268:**18372–18381.
28. Luka, S., J. Sanjuan, R. W. Carlson, and G. Stacey. 1993. *nolMNO* genes of *Bradyrhizobium japonicum* are co-transcribed with *nodYABCSUIJ,* and *nolO* Is in-

volved in the synthesis of the lipo-oligosaccharide nodulation signals. *J. Biol. Chem.* **268:**27053–27059.

29. Demont, N., F. Debellé, H. Aurelle, J. Dénarié, and J.-C. Promé. 1993. Role of the *Rhizobium meliloti nodF* and *nodE* genes in the biosynthesis of lipo-oligosaccharide nodulation factors. *J. Biol. Chem.* **268:**20134–20142.
30. Ramadas Bath, U., H. Mayer, A. Yokota, R. I. Hollingsworth, and R. W. Carlson. 1991. Occurrence of lipid A variants with 27-hydroxyoctasanoic acid in lipopolysaccharides from members of the family *Rhizobiaceae. J. Bacteriol.* **173:**2155–2159.
31. Demont, N., M. Ardourel, F. Maillet, D. Promé, M. Ferro, J.-C. Promé, and J. Dénarié. 1994. The *Rhizobium meliloti* regulatory *nodD3* and *syrM* genes control the synthesis of a particular class of nodulation factors *N*-acylated by (ω-1)-hydroxylated fatty acids. *EMBO J.* **13:**2139–2149.
32. Geiger, O., J. E. Thomas-Oates, J. Glushka, H. P. Spaink, and B. J. J. Lugtenberg. 1994. Phospholipids of *Rhizobium* contain *nodE* determined highly unsaturated fatty acid moeieties. *J. Biol. Chem.* **269:**11090–11097.
33. Ritsema, T., O. Geiger, P. Van Dillewijn, B. J. J. Lugtenberg, and H. P. Spaink. 1994. Serine residue 45 of nodulation protein NodF from *Rhizobium leguminosarum* bv. *viciae* is essential for its biological activity. *J. Bacteriol.* **176:**7740–7743.
34. Shearman, C. A., L. Rossen, A. W. B. Johnson, and J. A. Downie. 1986. The *Rhizobium leguminosarum* nodulation gene *nodF* encodes a polypeptide similar to acyl carrier protein and is regulated by *nodD* plus a factor in pea root exudate. *EMBO J.* **5:**647–652.
35. Bibb, M. J., S. Biro, H. Motamedi, J. F. Collins, and C. R. Hutchinson. 1989. Analysis of the nucleotide sequence of the *Streptomyces glaucescens tcmI* genes provide key information about the enzymology of polyketide antibiotic biosynthesis. *EMBO J.* **9:**2727–2736.
36. Hopwood, D. A., and D. H. Sherman. 1990. Molecular genetics of polyketides and its comparison to fatty acid biosynthesis. *Annu. Rev. Genet.* **24:**37–66.
37. Geiger, O., H. P. Spaink, and E. P. Kennedy. 1991. Isolation of the *Rhizobium leguminosarum* NodF nodulation protein: NodF carries a 4'-phosphopantetheine prostetic group. *J. Bacteriol.* **173:**2872–2878.
38. Debellé, F. and S. B. Sharma. 1986. Nucleotide sequence of *Rhizobium meliloti* RCR2011 genes involved in host specificity of nodulation. *Nucleic Acids Res.* **14:**7453–7472.
39. Sheldon, P. S., R. G. O. Kekwick, C. Sidebottom, C. G. Smith, and A. R. Slabas. 1990. 3-Oxoacyl-(acyl-carrier protein) reductase from avocado (*Persea americana*) fruit mesocarp. *Biochem. J.* **271:**713–720.
40. Roche, P., P. Lerouge, C. Ponthus, and J.-C. Promé. 1991. Structural determination of bacterial nodulation factors involved in the *Rhizobium meliloti*-alfalfa symbiosis. *J. Biol. Chem.* **266:**10933–10940.
41. Downie, J. A. 1989. The *nodL* gene from *Rhizobium leguminosarum* is homologous to the acetyl transferases encoded by *lacA* and *cysE. Mol. Microbiol.* **3:**1649–1651.

42. Evans, I. J., R. Jones, P. R. Woodley, J. R. Wilborn, and R. L. Robson. 1991. Nucleotide sequence and genetic analysis of the *Azotobacter chroococcum nifUSVWZM* gene cluster, including a new gene (*nifP*) which encodes a serine acetyl transferase. *J. Bacteriol.* **173:**5457–5469.
43. Baev, N. and A. Kondorosi. 1992. Nucleotide sequence of the *Rhizobium meliloti nodL* gene located in locus n5 of the *nod* regulon. *Plant Mol. Biol.* **18:**843–846.
44. Bloemberg, G. V., J. E. Thomas-Oates, B. J. J. Lugtenberg, and H. P. Spaink. 1994. Nodulation protein NodL of *Rhizobium leguminosarum O*-acetylates lipo-oligosaccharides, chitin fragments and *N*-acetylglucosamine *in vitro. Mol. Microbiol.* **11:**793–804.
45. Bloemberg, G. V., R. Lagas, G. van der Marel, J. H. van Boom, B. J. J. Lugtenberg, and H. P. Spaink. 1995 Substrate specificity and kinetic studies of nodulation protein NodL of *Rhizobium leguminosarum*. Biochem. **34:**12712–12720.
46. Firmin, J. L., K. E. Wilson, R. W. Carlson, A. E. Davies, and J. A. Downie. 1993. Resistance to nodulation of cv. Afghanistan pea is overcome by *nodX*, which mediates an *O*-acetylation of the *Rhizobium leguminosarum* lipo-oligosaccharide nodulation factor. *Mol. Microbiol.* **10:**351–360.
47. Clark, C. A., J. Beltrame, and P. A. Manning. 1991. The *oac* gene encoding a lipopolysaccharide *O*-antigen acetylase maps adjacent to the integrase-encoding gene on the genome of *Shigella flexneri* bacteriophage SF6. *Gene* **106:**43–52.
48. Roche, P., F. Debellé, F. Maillet, P. Lerouge, C. Faucher, G. Truchet, J. Dénarié, and J.-C. Promé. 1991. Molecular basis of symbiotic host specificity in *Rhizobium meliloti*: *nodH* and *nodPQ* genes encode the sulfation of lipo-oligosaccharide signals. *Cell* **67:**1131–1143.
49. Schwedock, J. and S. R. Long. 1990. ATP sulphurylase activity of the *nodP* and *nodQ* gene products of *Rhizobium meliloti. Nature* **348:**644–647.
50. Schwedock, J. S., C. X. Liu, T. S. Leyh, and S. R. Long. 1994. *Rhizobium meliloti* NodP and NodQ form a multifunctional sulfate-activating complex requiring GTP for activity. *J. Bacteriol.* **176:**7055–7064.
51. Schultze, M., C. Staehelin, H. Röhrig, M. John, J. Schmidt, E. Kondorosi, J. Schell, and A. Kondorosi. 1995 *In vitro* sulfotransferase activity of *Rhizobium meliloti* NodH protein: lipo chitooligosaccharide nodulation signals are sulfated after synthesis of the core structure. *Proc. Natl. Acad. Sci. USA,* **92:**2706–2709.
52. Geelen, D., P. Mergaert, R. A. Geremia, S. Goormachtig, M. Van Montagu and M. Holsters. 1993. Identification of *nodSUIJ* genes in Nod locus *1* of *Azorhizobium caulinodans*: evidence that *nodS* encodes a methyltransferase involved in Nod factor modification. *Mol. Microbiol.* **9:**145–154.
53. Stacey, G., S. Luka, J. Sanjuan, Z. Banfalvi, A. J. Nieuwkoop, J. Y. Chun, L. S. Forsberg, and R. W. Carlson. 1994. *nodZ*, a unique host-specific nodulation gene, is involved in the fucosylation of the lipooligosaccharide nodulation signal of *Bradyrhizobium japonicum. J. Bacteriol.* **176:**620–633.
54. Evans, I. J. and J. A. Downie. 1986. The *nodI* gene product of *Rhizobium leguminosarm* is closely related to ATP-binding bacterial transport proteins; nucleotide sequence analysis of the *nodIJ* genes. *Gene* **43:**95–101.

55. Fath, M. J. and R. Kolter. 1993. ABC transporters: Bacterial exporters. *Microbiol. Rev.* **57**:995–1017.

56. Vázquez, M., O. Santana, and C. Quinto. 1993. The NodI and NodJ proteins from *Rhizobium* and *Bradyrhizobium* strains are similar to capsular polysaccharide secretion proteins from Gram-negative bacteria. *Mol. Microbiol.* **8**:369–377.

57. Sutton, J. M., R. Rivilla, A. E. Davies, E. J. A. Lea, S. Ghelani, C. Finnie, G. Dean, and J. A. Downie. 1994. Functional analysis of NodO and NodT from *Rhizobium leguminosarum* biovar *viciae*. In *Advances in Molecular Genetics of Plant-Microbe Interactions, Vol.3*, eds. M. J. Daniels, J. A. Downie, and A. E. Osbourn, pp. 103–107. Kluwer Academic Publishers, Dordrecht.

58. Schlaman, H. R. M., R. J. H. Okker, and B. J. J. Lugtenberg. 1990. Subcellular localization of the *Rhizobium leguminosarum nodI* gene product. *J. Bacteriol.* **172**:5486–5489.

59. McKay, I. A. and M. A. Djordjevic. 1993. Production and excretion of Nod metabolites by *Rhizobium leguminosarum* bv. *trifolii* are disrupted by the same environmental factors that reduce nodulation in the field. *Appl. Env. Microbiol.* **59**:3385–3392.

60. Spaink, H. P., A. H. M. Wijfjes, and B. J. J. Lugtenberg. 1995 Rhizobium NodI and NodJ proteins determine the efficiency of secretion of lipo-chitin oligosaccharides. J. Bacteriol. **177**:6276–6281.

61. Rivilla, R. and J. A. Downie. 1994. Identification of a *Rhizobium leguminosarum* gene homologous to nodT but located outside the symbiotic plasmid. *Gene* **144**:87–91.

62. Ehrhardt, D. W., E. M. Atkinson, and S. R. Long. 1992. Depolarization of alfalfa root hair membrane potential by *Rhizobium meliloti*. *Science* **256**:998–1000.

63. Allen, N. S., M. N. Bennett, D. N. Cox, A. Shipley, D. W. Ehrhardt, and S. R. Long. 1994. Effects of Nod factors on alfalfa root hair Ca^{2+} and H^+ currents and on cytoskeletal behaviour. In *Advances in Molecular Genetics of Plant-Microbe Interactions,* Vol. 3, eds. M. J. Daniels, J. A. Downie, and A. E. Osbourn, pp. 107–113. Kluwer Academic Publishers, Dordrecht.

64. Price, N. P. J., B. Relić, F. Talmont, A. Lewin, D. Promé, S. G. Pueppke, F. Maillet, J. Dénarié, J.-C. Promé, and W. J. Broughton. 1992. Broad-host-range *Rhizobium* species strain NGR234 secretes a family of carbamoylated, and fucosylated, nodulation signals that are *O*-acetylated or sulphated. *Mol. Microbiol.* **6**:3575–3584.

65. Sanjuan, J., R. W. Carlson, H. P. Spaink, U. R. Bhat, W. M. Barbour, J. Glushka, and G. Stacey. 1992. A 2-*O*-methylfucose moiety is present in the lipo-oligosaccharide nodulation signal of *Bradyrhizobium japonicum*. *Proc. Natl. Acad. Sci. USA* **89**:8789–8793.

66. Mergaert, P., M. Van Montagu, J. Promé, and M. Holsters. 1993. Three unusual modifications, a D-arabinosyl, an N-methyl, and a carbamoyl group, are present on the Nod factors of *Azorhizobium caulinodans* strain ORS571. *Proc. Natl. Acad. Sci. USA* **90**:1551–1555.

67. Relić, B., F. Talmont, J. Kopcinska, W. Golinowski, J.-C. Promé, and W. J. Broughton. 1993. Biological activity of *Rhizobium* sp. NGR234 Nod-Factors on *Macroptilium atropurpureum*. *Mol. Plant- Microbe. Interact.* **6**:764–774.

68. López-Lara, I. M., J. D. J. Van den Berg, J. E. Thomas-Oates, J. Glushka, B. J. J. Lugtenberg, and H. P. Spaink. 1995. Structural identification of the lipo-chitin oligosaccharide nodulation signals from *Rhizobium loti*. *Mol. Microbiol.* **15:**627–638.

69. Heidstra, R., R. Geurts, H. Franssen, H. P. Spaink, A. van Kammen, and T. Bisseling. 1994. Root hair deformation activity of nodulation factors and their fate on *Vicia sativa*. *Plant Physiol.* **105:**787–797.

70. Van Brussel, A. A. N., R. Bakhuizen, P. C. Van Spronsen, H. P. Spaink, T. Tak, B. J. J. Lugtenberg, and J. W. Kijne. 1992. Induction of pre-infection thread structures in the leguminous host plant by mitogenic lipo-oligosaccharides of *Rhizobium*. *Science* **257:**70–72.

71. Truchet, G., P. Roche, P. Lerouge, J. Vasse, S. Camut, F. De Billy, J.-C. Promé, and J. Dénarié. 1991. Sulphated lipo-oligosaccharide signals of *Rhizobium meliloti* elicit root nodule organogenesis in alfalfa. *Nature* **351:**670–673.

72. Stokkermans, T. J. W. and N. K. Peters. 1994. *Bradyrhizobium elkanii* lipo-oligosaccharide signals induce complete nodule structures on *Glycine soja* Siebold et Zucc. *Planta* **193:**413–420.

73. Cárdenas, L., J. Domínguez, C. Quinto, I. M. López-Lara, B. J. J. Lugtenberg, H. P. Spaink, G. J. Rademaker, J. Haverkamp, and J. E. Thomas-Oates. 1995 Isolation, chemical structures and biological activity of the lipo-oligosaccharide nodulation signals from *Rhizobium etli*. Plant Mol. Biol. **29:**453–464.

74. Scheres, B., C. van de Wiel, A. Zalensky, B. Horvath, H. P. Spaink, H. Van Eck, F. Zwartkruis, A.-M. Wolters, T. Gloudemans, A. van Kammen, and T. Bisseling. 1990. The ENOD12 gene product is involved in the infection process during the pea-*Rhizobium* interaction. *Cell* **60:**281–294.

75. Horvath, B., R. Heidstra, M. Lados, M. Moerman, H. P. Spaink, J.-C. Promé, A. van Kammen, and T. Bisseling. 1993. Induction of pea early nodulin expression by Nod factors of *Rhizobium*. *Plant J.* **4:**727–733.

76. Pichon, M., E. P. Journet, A. Dedieu, F. de Billy, G. Truchet, and D. G. Barker. 1993. *Rhizobium meliloti* elicits transient expression of the early nodulin gene *ENOD12* in the differentiating root epidermis of transgenic afalfa. *Plant Cell* **40:**1199–1211.

77. Vijm, I., L. das Neves, A. van Kammen, H. Franssen, and T. Bisseling. 1993. Nod factors and nodulation in plants. *Science* **260:**1764–1765.

78. Kondorosi, E., P. Bauer, S. Brown, M. Crespi, T. Coba de la Pena, D. Dudits, J. El Turk, R. Esnault, E. Jurkevitch, Z. Magyar, M. Poiret, A. Savouré, R. Sallaud, C. Staehalin, M. Schultze, and A. Kondorosi. 1994. Effects of Nod factors on plants. In *Advances in Molecular Genetics of Plant-Microbe Interactions, Vol. 3*, eds. M. J. Daniels, J. A. Downie, and A. E. Osbourn, pp. 115–122. Kluwer Academic Publishers, Dordrecht.

79. Journet, E. P., M. Pichon, A. Dedieu, F. De Billy, G. Truchet, and D. G. Barker. 1994. *Rhizobium meliloti* Nod factors elicit cell-specific transcription of the ENOD12 gene in transgenic alfalfa. *Plant J.* **6:**241–249.

80. Recourt, K., J. Schripsema, J. W. Kijne, A. A. N. Van Brussel, and B. J. J. Lugtenberg. 1991. Inoculation of *Vicia sativa* subsp. *nigra* roots with *R. leguminosarum* bv.

viciae results in release of *nod* gene activating flavanones and chalcones. *Plant Mol. Biol.* **16**:841–852.

81. Estabrook, E. M. and C. Sengupta-Gopalan. 1990. Differential expression of phenylalanine ammonia-lyase and chalcone synthase during soybean nodule development. *Plant Cell* **3**:299–308.
82. Recourt, K., A. J. van Tunen, L. A. Mur, A. A. N. Van Brussel, B. J. J. Lugtenberg, and J. W. Kijne. 1992. Activation of flavonoid biosynthesis in roots of *Vicia sativa* subsp. *nigra* by inoculation with *Rhizobium leguminosarum* biovar *viciae*. *Plant Mol. Biol.* **19**:411–420.
83. Van Brussel, A. A. N., K. Recourt, E. Pees, H. P. Spaink, T. Tak, C.A. Wijffelman, J. W. Kijne, and B. J. J. Lugtenberg. 1990. A biovar-specific signal of *Rhizobium leguminosarum* bv. viciae induces increased nodulation gene-inducing activity in root exudate of *Vicia sativa* subsp. *nigra*. *J. Bacteriol.* **172**:5394–5401.
84. Schmidt, P. E., W. J. Broughton, and D. Werner. 1994. Nod factors of *Bradyrhizobium japonicum* and *Rhizobium* sp. NGR234 induce flavonoid accumulation in soybean root exudate. *Mol. Plant-Microbe Interact.* **7**:384–390.
85. Van Brussel, A. A. N., T. Tak, A. H. M. Wijfjes, H. P. Spaink, G. Smit, C. L. Díaz, and J. W. Kijne. 1993. Jamming of nodulation of *Vicia sativa* ssp. *nigra* by Nod factors of *Rhizobium leguminosarum* bv. *viciae*. In *New Horizons in Nitrogen Fixation*, eds. R. Palacios, J. Mora, and W. E. Newton, p. 261. Kluwer Academic Publishers, Dordrecht.
86. Relić, B., X. Perret, M. T. Estrada-García, J. Kopcinska, W. Golinowski, H. B. Krishnan, S. G. Pueppke, and W. J. Broughton. 1994. Nod factors of *Rhizobium* are a key to the legume door. *Mol. Microbiol.* **13**:171–178.
87. Savouré, A., Z. Magyar, M. Pierre, S. Brown, M. Schultze, D. Dudits, A. Kondorosi, and E. Kondorosi. 1994. Activation of the cell cycle machinery and the isoflavonoid biosynthesis pathway by active *Rhizobium meliloti* nod signal molecules in *Medicago* microcallus suspensions. *EMBO J.* **13**:1093–1102.
88. Yamada, A N., Shibuya, O. Kodama, and T. Akatsuka. 1993. Induction of phytoalexin formation in suspension-cultured rice cells by *N*-acetylchitooligosaccharides. *Biosci. Biotech. Biochem.* **57**:405–409.
89. Staehelin, C., J. Granado, J. Muller, A. Wiemken, R. B. Mellor, G. Felix, M. Regenass, W. J. Broughton, and T. Boller. 1994. Perception of *Rhizobium* nodulation factors by tomato cells and inactivation by root chitinases. *Proc. Natl. Acad. Sci. USA* **91**:2196–2200.
90. Felix, G., M. Regenass, and T. Boller. 1993. Specific perception of subnanomolar concentrations of chitin fragments by tomato cells: induction of extracellular alkalanization, changes in protein phosphorylation, and establisment of a refractory state. *Plant J.* **4**:307–316.
91. Debellé, F., C. Rosenberg, J. Vasse, F. Maillet, E. Martínez, J. Dénarié, and G. Truchet. 1986. Assigment of symbiotic developmental phenotypes to common and specific nodulation (*nod*) genetic loci of *Rhizobium meliloti*. *J. Bacteriol.* **168**:1075–1086.
92. Horvath, B., E. Kondorosi, M. John, J. Schmidt, I. Török, Z. Györgypal, I. Barabas, U. Wieneke, J. Schell, and A. Kondorosi. 1986. Organization, structure and symbiotic

function of *Rhizobium meliloti* nodulation genes determining host specificity for alfalfa. *Cell* **46:**335–343.

93. Faucher, C., F. Maillet, J. Vasse, C. Rosenberg, A. A. N. Van Brussel, G. Truchet, and J. Dénarié. 1988. *Rhizobium meliloti* host range *nodH* determines production of an alfalfa-specific extracellular signal. *J. Bacteriol.* **170:**5489–5499.
94. Cervantes, E., S. B. Sharma, F. Maillet, J. Vasse, G. Truchet, and C. Rosenberg. 1989. The *Rhizobium meliloti* host range *nodQ* encodes a protein which shares homology with translation elongation and initiation factors. *Mol. Microbiol.* **3:**745–755.
95. Faucher, C., S. Camut, J. Dénarié, and G. Truchet. 1989. The *nodH* and *nodQ* host range genes of *Rhizobium meliloti* behave as avirulence genes in *R. leguminosarum* bv. *viciae* and determine changes in the production of plant-specific extracellular signals. *Mol. Plant-Microbe Interact.* **2**:291–300.
96. Ardourel, M., N. Demont, F. D. Debellé, F. Maillet, F. De Billy, J.-C. Promé, J. Dénarié, and G. Truchet. 1994. *Rhizobium meliloti* lipo-oligosaccharide nodulation factors: Different structural requirements for bacterial entry into target root hair cells and induction of plant symbiotic developmental responses. *Plant Cell* **6:**1357–1374.
97. De Maagd, R. A., A. H. M. Wijfjes, H. P. Spaink, J. E. Ruiz-Sainz, C. A. Wijffelman, R. J. H. Okker, and B. J .J. Lugtenberg. 1989. *nodO*, a new *nod* gene of the *Rhizobium leguminosarum* biovar *viciae* Sym plasmid pRL1JI, encodes a secreted protein. *J. Bacteriol.* **171:**6764–6770.
98. Sutton, J. M., E. J. A. Lea, and J. A. Downie. 1994. The nodulation-signaling protein NodO from *Rhizobium leguminosarum* biovar *viciae* forms ion channels in membranes. *Proc. Natl. Acad. Sci. USA* **91:**9990–9994.
99. Sautter, C., G. V. Bloemberg, J. W. Kijne, B. J. J. Lugtenberg, I. Potrykus, and H. P. Spaink. Chitin oligosaccharides delivered by ballistic microtargeting can induce root nodule organogenesis, submitted.
100. Spaink, H. P., G. V. Bloemberg, A. H. M. Wijfjes, T. Ritsema, O. Geiger, I. M. López-Lara, M. Harteveld, D. Kafetzopoulos, A. A. N. Van Brussel, J. W. Kijne, B. J. J. Lugtenberg, K. M. G. M. Van der Drift, J. E. Thomas-Oates, I. Potrykus, and C. Sautter. 1994. The molecular basis of host-specificity in the *Rhizobium leguminosarum*-plant interaction. In *Advances in Molecular Genetics of Plant-Microbe Interactions, Vol.3*, eds. M. J. Daniels, J. A. Downie, and A. E. Osbourn, pp. 91–98. Kluwer Academic Publishers, Dordrecht.
101. Lie, T. A. 1971. Symbiotic nitrogen fixation under stress conditions. *Plant and Soil* **31:**117–127.
102. Davis, E. O., I. J. Evans, and A. W. B. Johnston. 1988. Identification of *nodX*, a gene that allows *Rhizobium leguminosarum* biovar *viciae* strain TOM to nodulate Afghanistan peas. *Mol. Gen. Genet.* **212:**531–535.
103. Lewin, A., E. Cervantes, W. Chee-Hoong, and W. J. Broughton. 1990. *nodSU*, two new *nod* genes of the broad host range *Rhizobium* strain NGR234 encode host-specific nodulation of the tropical tree *Leucaena leucocephala*. *Mol. Plant-Microbe Interact.* **3:**317–326.

104. Krishnan, H. B., A. Lewin, R. Fellay, W. J. Broughton, and S. G. Pueppke. 1992. Differential expression of *nodS* accounts for the varied abilities of *Rhizobium fredii* USDA257 and *Rhizobium* sp. strain NGR234 to nodulate *Leucaena* spp. *Mol. Microbiol.* **6:**3321–3330.

105. Staehelin, C., M. Schultze, E. Kondorosi, R. B. Mellor, T. Boller, and A. Kondorosi. 1994. Structural modifications in *Rhizobium meliloti* nod factors influence their stability against hydrolysis by root chitinases. *Plant J.* **5:**319–330.

106. Bono, J.-J., J. Riond, K. C. Nicolaou, N. J. Bockovich, V. A. Estevez, J. V. Cullimore, and R. Ranjeva. 1995. Characterization of a binding site for chemically synthesized lipo-oligosaccharide NodRm factors in particulate fractions prepared from roots. *Plant J.* **7:**253–260.

107. Shibuya, N., H. Kaku, K. Kuchitsu, and M. J. Maliarik. 1993. Identification of a novel high-affinity binding site for *N*-acetylchitooligosaccharide elicitor in the membrane fraction from suspension-cultured rice cells. *FEBS Lett.* **329:**75–78.

108. Baureithel, K., G. Felix, and T. Boller. 1994. Specific, high affinity binding of chitin fragments to tomato cells and membranes: Competitive inhibition of binding by derivatives of chitooligosaccharides and a Nod factor of *Rhizobium. J. Biol. Chem.* **269:**17931–17938.

109. Schmidt, J., H. Röhrig, M. John, U. Wieneke, G. Stacey, C. Koncz, and J. Schell. 1993. Alteration in plant growth and development by *Rhizobium nodA* and *nodB* genes involved in the synthesis of oligosaccharide signal molecules. *Plant J.* **4:**651–658.

110. De Jong, A. J., J. Cordewener, F. Lo Schiavo, M. Terzi, J. Vandekerckhove, A. van Kammen, and S. C. de Vries. 1992. A carrot somatic embryo is rescued by chitinase. *Plant Cell* **4:**425–433.

111. De Jong, A. J., R. Heidstra, H. P. Spaink, M. V. Hartog, T. Hendriks, F. Lo Schavio, M. Terzi, T. Bisseling, A. van Kammen, and S. C. de Vries. 1993. A plant somatic embryo mutant is rescued by rhizobial lipo-oligosaccharides. *Plant Cell* **5:**615–620.

112. Truchet, G., D. G. Barker, S. Camut, F. De Billy, J. Vasse, and T. Huguet. 1989. Alfalfa nodulation in the absence of *Rhizobium, Mol. Gen. Genet.* **219:**65–68.

113. Collinge, D. B., K. M. Kragh, J. D. Mikkelsen, K. K. Nielsen, U. Rasmussen, and K. Vad. 1993. Plant chitinases. *Plant J.* **3:**31–40.

114. Benhamou, N. and A. Asseling. 1993. Attempted localization of a substrate for chitinases in plant cells reveals abundant *N*-acetyl-D-glucosamine residues in secondary walls. *Biol. Cell* **67:**341–350.

115. Spaink, H. P., A. Aarts, G. V. Bloemberg, J. Folch, O. Geiger, H. R. M. Schlaman, J. E. Thomas-Oates, K. Van de Sande, P. Van Spronsen, A. A. N. Van Brussel, A. H. M. Wijfjes, and B. J. J. Lugtenberg. 1993. Rhizobial lipo-oligosaccharides: their biosynthesis and their role in the plant. In *Advances in Molecular Genetics of Plant-Microbe Interactions,* Vol. 2, eds. E. Nester and D. P. P. S. Verma, pp. 151–162. Kluwer Academic Publishers, Dordrecht.

116. Spaink, H. P., A. H. M. Wijfjes, T. B. Van Vliet, J. W. Kijne, and B. J. J. Lugtenberg. 1993. Rhizobial lipo-oligosaccharide signals and their role in plant morphogenesis:

are analogous lipophilic chitin derivatives produced by the plant? *Austr. J. Plant Physiol.* **20:**381–392.

117. Wagner, G. P., J. Lo, R. Laine, and M. Almeder. 1993. Chitin in the epidermal cuticle of a vertebrate (*Paralipophrys trigloides*, Blenniidae, Teleostei). *Experientia* **49:**317–319.

118. Rosa, F., T. D. Sargent, M. L. Rebbert, G. S. Michaels, M. Jamrich, H. Grunz, E. Jonas, J. A. Winkles, and I. B. David. 1988. Accumulation and decay of DG42 gene products follow a gradient pattern during *Xenopus* embryogenesis. *Devel. Biol.* **129:**114–123.

119. Jordan, D. C. 1982. Transfer of *Rhizobium japonicum*, Buchanan 1980 to *Bradyrhizobium* gen. nov., a genus of slow growing, root nodule bacteria from leguminous plants. *Int. J. Syst. Bacteriol.* **32:**136–139.

120. Martínez-Romero, E., L. F. M. Segovia, A. A. Franco, P. Graham, and M. A. Pardo. 1991. *Rhizobium tropici*, a novel species nodulating *Phaseolus vulgaris* L. beans and *Leucaena* sp. trees. *Int. J. Syst. Bacteriol.* **41:**417–426.

121. Herrera, M. A., E. J. Bedmar, and J. Olivares. 1985. Host specificity of *Rhizobium* strains from nitrogen-fixing trees and nitrogenase activities of strain GRH2 in symbiosis with *Prosopis chilensis. Plant Science* **42:**177–182.

122. Elkan, G. H. 1992. Taxonomy of the rhizobia. *Can. J. Microbiol.* **38:**446–450.

123. Dreyfus, B. L. and Y. P. Dommergues. 1981. Nitrogen-fixing nodules induced by *Rhizobium* on the stem of the tropical legume *Sesbania rostrata. FEMS Microbiol. Lett.* **10:**313–317.

124. Poupot, R., E. Martínez-Romero, and J.-C. Promé. 1993. Nodulation factors from *Rhizobium tropici* are sulfated or non-sulfated chitopentasaccharides containing an *N*-methyl-*N*-acylglucosaminyl terminus. *Biochemistry* **32:**10430– 10435.

CHAPTER

2

Metabolism and Compartmentation of Carbon and Nitrogen in Legume Nodules

S. Tajima and H. Kouchi

INTRODUCTION

Symbiotic association of legume plants with soil bacteria of the genera *Rhizobium, Bradyrhizobium,* and *Azorhizobium* leads to the formation of nodules in which the microsymbionts are capable of fixing atmospheric dinitrogen. Symbiotic bacteria are fully dependent for their respiration and nitrogen fixation on the supply of photosynthate from the host plants. The nitrogen fixed in the bacteroids is exported to the plant cell cytosol as ammonium and assimilated to specific forms for translocation to the plant shoots. The nodule is a specialized symbiotic organ in which both the plant cell and microsymbiont develop specific structures and metabolic functions required for nitrogen fixation.

Recent advances in molecular studies on the symbiotic interaction between legume plants and rhizobium bacteria allowed the identification of the signal molecules that trigger nodule organogenesis. The signal molecules, lipo-chitin oligosaccharides (Nod factors), produced by activation of bacterial *nod* genes, are able to elicit initial steps of nodule organogenesis in a host-specific manner[1,2] and, in some cases, induce genuine nodule structures.[3,4] However, it is not clear whether these structures elicited by Nod factors are accompanied by activation of the metabolic functions specific to symbiotic nodules.

The basic features of nodule metabolism have been extensively studied during the past 10 years.[5–7] For instance, there have been great advances in understanding metabolite exchange between plant cell cytoplasm and microsymbiont, compartmentation in relation to carbon and nitrogen metabolism, and regulation mechanisms of O_2 diffusion in nodule tissues.[8–14] Our knowledge about these aspects of nodule physiology, however, has been obtained mainly by the analysis of mature nodules, and it still remains to be elucidated how symbiotic metabolism is organized through the interactions of both partners of symbiosis during the early stages of nodule formation. In this paper, we summarize current knowledge of the major flow of carbon and nitrogen compounds in nitrogen-fixing nodules. In addition, we discuss the possible connection between gene expression and metabolic regulation to clarify the perspectives for future research.

COMPARTMENT AND METABOLIC FUNCTIONS IN NODULES

Nodule Structure and Compartment

Infection and Nodule Formation.

The nodule formation process involves several consecutive steps. In most legumes, rhizobia attach to growing root hairs and cause root hair deformation and/or curling. The bacteria then invade root hair cells through infection threads, which grow and penetrate the root cortical cells.[1,8,9] Concomitantly with this infection process, cortical cell division is induced and nodule primordia are formed. The bacteria are released into the nodule cortical cell cytoplasm from infection threads. They are surrounded by a membrane of plant origin (the symbiosome membrane, SM) and differentiate into bacteroids.[8,9] Vascular strands that connect the root vascular cylinder and nodule primordia are formed, and nodule primordia differentiate into fully developed nodules in which the bacteroids fix atmospheric nitrogen.

There are two major types of nodules. Legumes belonging to *Medicago, Pisum, Trifolium, Sesbania, Oxytropis,* and *Astragalus* form "indeterminate" nodules which are initiated from inner cortical cells and have a persistent meristem.[5–7] Because of their continuing growth, they are generally cylindrical in shape, and different symbiotic stages are distributed spatially from the apical to basal part of the nodule. Species belonging to *Glycine, Phaseolus, Lotus, Arachis,* and *Vigna* form "determinate" nodules with a spherical shape. The meristem of determinate nodules is initiated from subepidermal cells and meristematic activity is limited in the early stages of nodule development. The mature nodules have no persistent meristem. These differences in nodule ontogeny are determined by the plant genotype and are independent of microsymbiont genome.

One of the most striking differences in metabolism between these two types of nodule is that amides (principally asparagine) are the major nitrogen compounds exported from indeterminate nodules, whereas ureides (allantoin and allantoic acid) are the major nitrogen compounds exported from determinate nodules.[5,7,12] However, *Arachis* and *Lotus* nodules are believed to be exceptions to this rule because they form determinate nodules but do not export ureides.[12,15]

Other features of the two types of nodules have been summarized by Sprent.[16] Despite this simple classification, there are various exceptions and intermediate types. In addition, numerous nodule-bearing legumes have not been studied.[17]

Symbiosome

Symbiosome is the most prominent compartment that is specific in nodules, although the similar structure is common in many endocytobiotic systems as a lysosome-like structure enclosing the parasite in the host cytoplasm.[18,19] The symbiosome provides an environment in which bacteroids can exist in the plant cell without inducing a serious defense reaction from plant.[20] Bacteroids are considered stable under certain chemical control by the plant cell via the activity and permeability of the SM. Figure 2-1 shows a characteristic profile of the SM. The characteristics of the SM have been studied intensively and have recently been

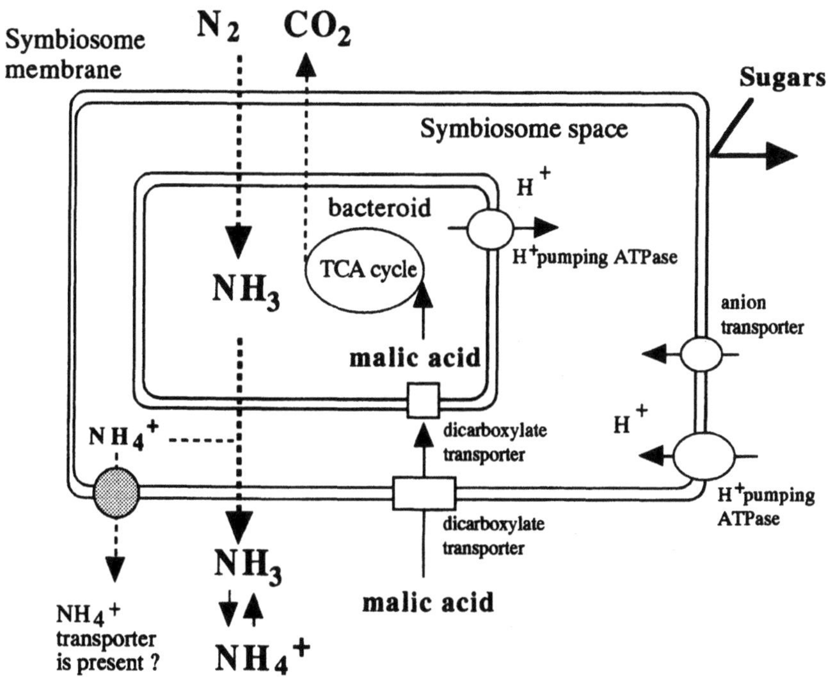

Figure 2-1. Proposed profile of metabolite exchange through symbiosome membrane.

reviewed by Day and Udvardi.[19] The SM is derived from plant plasma membrane[21] but modified specifically to endosymbiosis. For instance, the SM contains a number of nodule-specific proteins (nodulins) as major constituents (see the section "Nodulin genes and metabolic regulation" for details).

The transport functions of the SM have been most intensively studied. It has been well established that the SM has a dicarboxylate transporter, which plays a key role in supplying C_4-dicarboxylates to bacteroids.[21–23] The SM dicarboxylate transporter has been well characterized biochemically but not yet isolated. The dicarboxylate transport activity of the SM is affected by phosphorylation of the SM,[24] and a Ca^{2+}-dependent protein kinase was identified on the SM.[25,26] The presence of a H^+-ATPase of the host plasma membrane type has been also demonstrated.[27] In contrast to the presence of active dicarboxylate transporter, the SM appears to be impermeable to a range of sugars. The one exception is the SM from french bean, which has been reported to transport sugars.[28]

The space between the bacteroid and the SM is called the *symbiosome space.* Since H^+-pumping mediated by a membrane ATP synthase is operative on both host and symbionts' membranes,[29] the symbiosome space is proposed to be acidic.[30,31] Day and Udvardi[31] have reported that the pH difference between the bacteroid and symbiosome space is 1.5. Haaker et al.[32] suggested, however, that such an acidic condition of the symbiosome space is not relevant because nitrogenase activity of bacteroids in the symbiosome was inhibited when the pH value in the symbiosome space is acidic or above 8. A number of enzymes have been identified in the symbiosome space, e.g., acid phosphatase, α-mannosidase, protease, and trehalase.[13] The physiological significance of these enzymes in the symbiosome space is totally unknown.

Organogenesis of the symbiosome is dependent on the presence of intercellular bacteria. Some mutant strains of *B. japonicum*, which form ineffective nodules on soybean, have been shown to result in disintegration of the SM at early stages of nodule growth or significant alterations of the protein components of the SM.[9,33–35] A putative signal peptidase secreted from the bacteroids might be responsible for modifying the SM proteins to maintain the stability of the SM.[36] The SM appears to be the first target of proteolysis during nodule senescence.[37,38] The mechanisms of the interactions of the host cell cytoplasm and microsymbiont in the formation, structure, and stability of the SM are largely unknown.

Bacteroids cease cell division after the symbiosome has matured. However, they are able to retain proliferating activity ex planta when the isolated bacteroids are incubated on an agar medium containing a very low concentration of yeast extract.[39] Polyamines in the medium have been indicated to affect the proliferating activity of the isolated bacteroids.[40] From these results, it has been suggested that as rhizobia differentiate into bacteroids they become sensitive to some chemicals, presumably polyamines from the plant cell. In other cases, isoflavonoids are known to be strong growth inhibitors of the proliferation of *Bradyrhizobium japonicum*[40] or *R. loti*[41] cell division, even though they are basically *nod*-gene activators released from the plant.

Metabolic Environment

Microaerobiosis

A major factor affecting nodule metabolism is the extremely low free O_2 concentration in the infected cell. Studies with O_2 microelectrodes[42,43] have shown that the O_2 concentration in the central zone of the nodule is below the detection limit for the electrodes (i.e., less than ca. 200 nM O_2). Using a spectroscopic technique, the O_2 concentration in nodules has been shown to range from about 5 to 50 nM O_2.[44,45]

The nitrogenase system is O_2-labile, and the low oxygen condition protects the enzyme from O_2 inactivation. However nitrogenase needs a large amount of ATP for the enzyme reaction. This range of O_2 concentration is not far from the Km(O_2) for the bacteroid terminal oxidase enzymes (5–20 nM),[46] but it is lower than that for the terminal oxidase of nodule mitochondria (50–100 nM).[47] Thus, the capability of maintaining high respiration under low oxygen levels is thought to play a central role in nodule metabolism for supplying a large amount of ATP for nitrogen fixation, active transport, ammonia assimilation, and amide/ureide biosynthesis.

To maintain a high flux of O_2 at this low O_2 concentration, the cytoplasm of the infected cells have a high concentration (ca. 3 mM) of a myoglobin-like protein called leghemoglobin (Lb).[48,49] Lb can reversibly bind to O_2 and facilitate the diffusion of O_2 from the surface of the infected cell to the bacteroids found within. In intact nodules being in equilibrium with air, 16–42% of the Lb is oxygenated (equivalent to an oxygen concentration of 9–35 nM).[50]

The biochemical process for maintaining Lb function is still unknown. Ferric leghemoglobin reductase may be important in keeping the oxidative status of Lb to Lb^{2+} (active form) because phenolic compounds or nitrite[51] can be accumulated in nodules and these comounds can oxidize Lb^{2+} to Lb^{3+}. This enzyme has been purified and cloned from a cDNA library of soybean nodules.[52,53] The gene was found to be expressed in various tissues, but high expression was only found in nodules and leaves, suggesting the enzyme is not specific to Lb. In addition, high temperature,[51] the formation of nitrosyl- leghemoglobin after nitrate application,[54] and drought stress[55,56] can also influence the redox state of Lb.

Even in microaerobic conditions, an adenylate energy charge, NADH/NAD ratio, and NADPH/NADP ratio values were found to be similar to a normal aerobic plant cell when whole nodule tissue was analyzed.[57–59] The nodule NADH/NAD ratio value of 0.08 was slightly lower than those of aerobic cells. The high nodule NADPH/NADP ratio value of 3.8 suggests that malic enzyme or the pentose phosphate shunt is active for generating NADPH.[57] Adenylate energy charge values of whole nodules tend to be lower (0.6–0.9) than those of aerobic cells. These metabolic parameters suggest that nodule cells in normal conditions can maintain an oxidative respiratory pathway under the low free oxygen conditions of the nodule interior.

Oxygen Barrier

To explain this anaerobic environment in the infected zone of nodules, the presence of an oxygen barrier surrounding the infected zone has been proposed.[42,43,60] The most interesting phenomena is that the oxygen barrier can adapt to the change of ambient O_2 concentrations to maintain O_2 level in the nodule-infected zone constant. When the nodules are exposed to 100% O_2 directly from 20% O_2, the nitrogenase activity is lost rapidly and irreversibly. However, when ambient pO_2 was increased gradually, pO_2 in the infected zone remained at an initial level and the nitrogenase was still active even under 100% ambient O_2 without significant increases in respiratory CO_2 evolution.[14,50,61] This can be explained as that the oxygen barrier could change the O_2 permeability in response to increase in the ambient pO_2. This adaptation is reversible. When the nodules once adapted to 100% O_2 were put in 20% normal O_2, the respiration and nitrogenase activity declined transiently and then returned to the initial levels. Thus, a variable barrier to O_2 diffusion in the nodule cortex has been proposed as a factor responsible for regulating the O_2 concentration in the central zone of the legume nodule.[44,48]

In addition, limitation of carbohydrate supply has been proposed to regulate the oxygen barrier. Treatments that disrupt or divert the supply of photosynthate to the nodules (e.g., stem girdling, nodule detachment, nitrate supply) reduced the infected cell O_2 concentration and nitrogenase activity of nodules.[62,63] According to these data, limitation of carbohydrate supply is proposed to cause O_2 limitation through a tighter oxygen barrier.[63]

A variety of hypotheses have been forwarded to account for how this barrier may operate. An extensive review has been published about this topic.[14] In general, these hypotheses offer mechanisms to explain how nodules may be able to alter the path of O_2 diffusion from gas-filled channels (high permeability pathways) to aqueous or protein-filled routes (low permeability pathways).

Further empirical studies are required to test the various hypotheses that have been formulated.

Carbon Metabolism and ATP Generation

Plant Carbon Metabolism

Figure 2-2 illustrates the overall carbon and nitrogen flow between various nodule compartments, i.e., vascular bundle, uninfected cell, infected cell, symbiosome, and bacteroids.

Legume nodules are a major sink in the root system. They consume about 12–17 g of carbohydrate ($C_6H_{12}O_6$) for each gram of fixed nitrogen, which covers

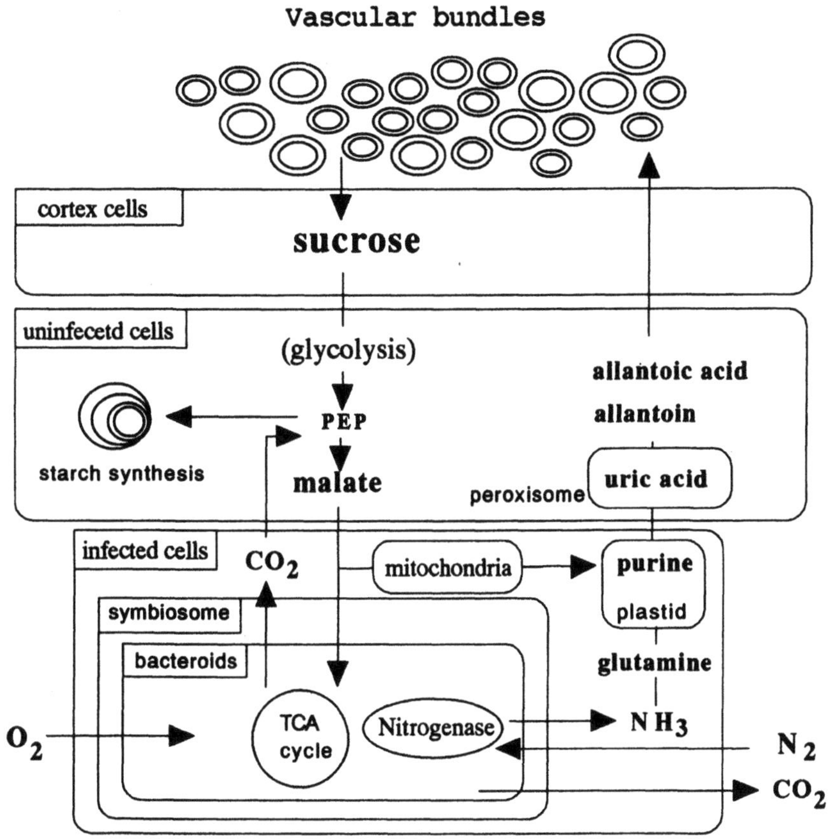

Figure 2-2. Supposed carbon and nitrogen metabolism in cortex cells, uninfected cells, and infected cells of soybean nodules.

the cost of the nitrogenase reaction and ammonia assimilation.[11,12] In other words, nodules respire, mostly in bacteroid metabolism, approximately 15–30% of the photosynthate imported to the whole root system, and an additional 30% of that photosynthate is used as carbon skeletons of nitrogenous compounds for translocation to the shoots.[7,11,12,64,65] The effect of carbon supply on nitrogen fixation has been extensively reviewed.[11] Because nodule nitrogenase activity is not depending on stored carbon in nodules,[66,67] a possible key factor in regulating nodule nitrogen fixation may be increasing a supply of current photosynthate. However, all attempts to increase nitrogen fixation by maximizing photosynthetic rate have been reported to meet no success.[68,69] In contrast, reduction of photosynthate supply is reportedly connected to rapid decline of nitrogenase activity and respiration rate.[70,71] Since a sucrose transporter protein was recently identified as the phloem-loading system of leaves,[72] the biochemical and genetic analysis on similar protein in nodules would be possible. These studies may elucidate the molecular mechanism of photosynthate supply to nodules.

Once unloaded, the sucrose is metabolized in the cells of the nodule cortex or in the infected and uninfected cells of the central zone (determinate nodules). Since the major form of reduced carbon for bacteroids is C4-dicarboxylates (see next section), the possible carbon flux to bacteroids is suggested to be primarily dependent on the conversion of sucrose to C_4-dicarboxylic acids (mainly malic acid).[11,12,65] In soybean nodules, three-dimensional organization of uninfected tissue was studied.[73,74] The data showed that every infected cell is in contact with at least one uninfected cell.[73] In addition, the uninfected cells often occur arranged in lines oriented approximately radially in the central region.[74] These results suggest a metabolic cooperativity between the two types of cell. It is not yet clear how the pathways for conversion of sucrose to C_4-dicarboxylic acids are differentiated between cortical, uninfected, and infected cells. At present, however, a number of the studies on the distribution of enzymes involved in sucrose catabolism suggest that cortical cells and uninfected cells have greater responsibility for sucrose breakdown rather than infected cells.[75,76]

The first enzyme to degrade sucrose supposedly is sucrose synthase (SS), which converts sucrose to uridine-5'-diphosphoglucose (UDPG). Although SS and alkaline invertase are both present in high concentrations in nodules,[75] SS was suggested to be a major pathway for the primary breakdown of sucrose. A nodule-specific form of SS (nodulin 100) has been identified in soybean nodules.[77] The SS is a very abundant protein in nodules, and an immunohistochemical study indicated that it is localized predominantly in uninfected cells and the inner cortex of soybean nodules.[76]

Enzymes for organic acid interconversion and dark CO_2 fixation, phosphoenolpyruvate carboxylase (PEPC) or carbonic anhydrase, have also been detected in nodules at high concentration.[78,79] These enzymes are important not only for supporting respiration of bacteroids, but also for supplying carbon skeleton for synthesizing amides and ureides for translocation of fixed nitrogen to the shoots. Among these enzymes, PEPC is believed to play a key role by converting phosphoenolpyruvate to oxaloacetic acid (thus bypassing pyruvate formation), which is transported into mitochondria and forms malic acid (most possible *in vivo* substrate of bacteroids) through malate dehydrogenase.[11,12]

Pyruvate decarboxylase (PDC) and alcohol dehydrogenase were reported to be present in soybean nodules.[80] However, no significant accumulation of ethanol or acetaldehyde has been observed in the intact nodules. Interestingly, decapitation of the shoot caused rapid elevation of ethanol and acetaldehyde levels in nodules. Since these compounds are volatile they can be easily lost from nodules by passive diffusion. The biochemical mechanism for how organic acids favor PEPC rather than PDC is unknown. The regulation of these pathways would be important in preventing the loss of respiratory substrates and is also a common feature in plants adapted to flooding.[81]

Cellular compartmentation of this metabolism in nodules has been analyzed by separating single cells or protoplasts of uninfected cells and infected cells from nodules. Although the values of enzyme activities detected in these cells were not fully reliable because inactivation occurring during isolation steps may differ for individual enzymes and cell fractions, many experiments have shown that the cortical cells and uninfected cells in nodules have very high activities of glycolytic enzymes responsible for the degradation of sucrose.[82–84]

Mitochondria isolated from cowpea nodules showed higher respiratory activity relative to those from root cells.[47,85] This activity is stimulated by adding leghemoglobin to the reaction mixture,[85,86] suggesting they are adaptive to microaerobic conditions. The high density of mitochondria found close to the plasma membrane in all infected cells also suggests an adaptation to microaerobic environment. Since mitochondria in the nodule infected zone are functioning under microaerobic conditions, they also might supply 2-oxo-glutarate for ammonia assimilation.[11,12]

Soybean nodules contain large amounts of cyclitols, such as *myo*-inositol, ononitol and pinitol.[87,88] These compounds are also found in other organs of soybean, but their concentrations are particularly high in nodules. However, their metabolic turnover rates are very low as indicated by ^{13}C-tracer experiments[64,65] and therefore their significance in carbon metabolism is still obscure. Very recently, it was proposed that these cyclitols may function to stabilize the SM.[89] Trehalose has also been found in high concentrations in all legume nodules and is known to be synthesized by the bacteroids.[11,12,64,90]

Bacterial Carbon Metabolism

The most efficient respiratory substrates for isolated bacteroids are C_4-dicarboxylic acids.[5,6,10–12] Efficient dicarboxylate absorption by the symbiosome has already been described.[10,12,19] Accumulated dicarboxylates in the symbiosome space can be actively transported into bacteroids by the dicarboxylate transporter on the bacteroid membrane.[10,12,19] The principal role of dicarboxylic acids in bacteroid respiration and nitrogen fixation is well documented from various viewpoints. The most conclusive evidence has come from studies of dct (dicarboxylate transporter) mutants. All mutants of *Rhizobium* and *Bradyrhizobium* sp. defective in dicarboxylate transport (Dct^-) can form nodules, but show little nitrogenase activity (Fix^-).[12,91,92] Since Dct^- bacteroids are likely to incorporate organic acids by passive diffusion, the mutant bacteroids' respiration is not able to support the large ATP demand of the nitrogenase system.[91] Mutants defective in enzymes for metabolizing organic acids also form Fix^- nodules.[12]

Isolated bacteroids do not utilize sugars as a reduced carbon source.[12,19] Such insensitivity to sugars seems to be a common characteristic of isolated bacteroids, and the activities of enzymes of glycolytic pathways are very low in bacteroids.[75] A pleiotropic mutant of *R. meliloti* that failed to use sugars as a carbon source because of the lack of an uptake system could still form Fix^+ nodules,[93] indicating that utilization of sugars is not essential for nitrogen fixation.

How is dicarboxylate oxidized? Tracer experiments (especially radiorespirometry) have shown that bacteroids have the capability to oxidize dicarboxylates very rapidly, mainly through the tricarboxylic acid (TCA) cycle[11,12,94,95,96] under microaerobic conditions when Lb is present in the reaction mixture. The degradation stopped immediately when O_2 supply was reduced. During the degradation there was no accumulation of pyruvate, acetate, or lactate, suggesting that respiration of isolated bacteroids is strictly aerobic.[95]

Isolated bacteroids can use amino acids as respiratory substrates.[11,12] *B. japonicum* bacteroids have a glutamate uptake system that stimulates respiration and nitrogenase activity. The significance of the β-aminobutyric acid pathway (GABA-shunt), which circumvents part of the TCA cycle, has been shown in *R. meliloti* bacteroid metabolism.[97] The pool size of GABA in *R. meliloti* bacteroids is extremely high compared to that in the free-living state.[98] Moreover, an *R. meliloti* mutant with reduced succinic semialdehyde dehydrogenase, which catalyzes the final step of the GABA-shunt, exhibits a significant reduction of symbiotic nitrogen fixation activity.[99] In contrast, glutamate comprises the dominant pool and the concentration of GABA is very low in *B. japonicum* bacteroids.[12,64] Salminen and Streeter[12,100] have proposed that the GABA-shunt does not operate in *B. japonicum* bacteroids because of a lack of glutamate decarboxylase activity, the first enzyme of the GABA-shunt. Therefore, the significance of the GABA-shunt to supply energy for the nitrogenase system may be limited to *Rhizobium* species.

The TCA cycle in bacteroids is easily inhibited by reductant accumulation. When bacteroids were incubated under microaerobic conditions without Lb or with slow shaking, the TCA cycle was stalled after a short incubation time and pyruvate accumulation was observed.[95] Decarboxylases, which are essential for TCA cycle operation in bacteroids (pyruvate dehydrogenase, malic enzyme, isocitrate dehydrogenase, and oxoglutarate dehydrogenase, etc.), are reported to be sensitive to accumulation of NAD(P)H,[101] and under microaerobic conditions NADH, NADPH or other reductants could tend to accumulate in the bacteroids, resulting in the diversion of the carbon flow to poly-β-hydroxybutyrate.[96]

How is acetyl-CoA formed from dicarboxylate to operate TCA cycle continuously?[5] Bacteroids appear to use malic enzymes for generating pyruvate from malate.[11,12,102,103] Other pyruvate-supplying enzymes have a much lower activity than the malic enzymes. In free-living conditions both NAD-type and NADP-type malic enzymes are present at similar levels in various *Rhizobium* and *Bradyrhizobium* species.[11,12] NADP-malic enzyme has a much higher affinity for L-malic acid ($Km = 0.16$ mM) than NAD-malic enzyme ($Km = 2.5$ mM). However, Driscoll et al.[103] reported that an NAD-malic enzyme-defective mutant of

Rhizobium meliloti was Nod⁺ but Fix⁻. This mutant still possessed low NADP-malic enzyme activity, suggesting that the NAD-malic enzyme is essential for nitrogen fixation by bacteroids. Finan et al.[104] also have succeeded in isolating an NADP-malic enzyme-defective mutant, and it formed Fix⁺ nodules. These data strongly suggest that only NAD-malic enzyme is essential for the bacteroids' nitrogenase activity and that the NADP-malic enzyme might play a different role, such as the survival of free-living cells in poor nutrient conditions. When the NAD-malic enzyme operates under high malate concentration in bacteroids,[103] it generates a large amount of CO_2 other than respiratory CO_2 evolution. High PEPC activity[78,79] may be important to maintain low CO_2 concentration in infected cell cytosol and recycle the carbon to dicarboxylic acid.

Purified malic enzyme activities are strongly influenced by the NAD(P)H/NAD(P) ratio. High NADPH/NADP ratio values in the bacteroids[57] have been found to be inhibitory for NADP-malic enzyme.[102] Day et al.[105] proposed that high levels of malate concentrations in the bacteroids could divert the carbon flow from TCA cycle to poly-β-hydroxybutyrate synthesis because accumulated NAD(P)H would inhibit TCA cycle operation. They observed that the addition of a relatively high concentration (0.5 mM) of malate to bacteroids under low O_2 concentration resulted in low nitrogenase activity and the carbon was diverted to poly-β-hydroxybutyrate. Tajima et al.[95] also observed pyruvate accumulation in the reaction mixture when isolated soybean bacteroids from soybean nodules were incubated with 0.5 mM malate.

In *R. meliloti* bacteroids NADP-malic enzyme activity is very low.[103,104] However, in *B. japonicum* bacteroids NAD- and NADP-malic enzyme activities have been detected at almost the same magnitude,[104] suggesting that in *B. japonicum*, the NADP-malic enzyme has a different metabolic role of that in *Rhizobium meliloti.* In *E. coli* these two enzymes are reported to have different physiological roles.[106]

The essential role of the NAD-malic enzyme and the TCA cycle operation suggest that oxidation of NADH, not NADPH, is coupling to electron transfer through the respiratory chain in bacteroids. *B. japonicum* bacteroids have been reported to have a respiratory terminal oxidase that is induced only in bacteroids and is essential to transfer O_2 from oxygenated Lb to the respiratory complex in the bacteroid cytoplasmic membrane.[107,108] This cytochrome oxidase system has been intensively surveyed by spectroscopic measurements.[107]

In addition to catabolic metabolism, bacteroids have the capability to form sugars from dicarboxylate[12,94,95] and synthesize novel exopolysaccharides, which can be excreted to the symbiosome space.[12] Streeter et al.[109] identified a novel exopolysaccharide synthesis by *Bradyrhizobium japonicum* bacteroids. *B. japonicum* (Group I) bacteroids produce an exopolysaccharide that is composed of galactose, rhamnose, and 2-o-methylglucuronic acid. In contrast, *B. elkanii* (formerly *B. japonicum* Group II)[109] bacteroids produce an exopolysaccharide composed of rhamnose and 4-o-methylglucuronic acid. These exopolysaccharides have been proposed to play a protective role for the bacterium during survival in the soil after nodule senescence and are predicted to possibly influence interstrain competitiveness of bacteria during infection.[110]

Nitrogen Metabolism

Bacterial nitrogen metabolism

The flow of fixed nitrogen to the shoots is of critical importance for the plant to benefit from symbiotic nitrogen fixation.

Nitrogenase in bacteroids catalyzes the conversion of atmospheric dinitrogen to ammonia using ATP and reductant.[5,7,111] Since the substrate specificity of nitrogenase is loose, various compounds are reported to be reduced, and as a result it causes waste of ATP. A by-product that is physiologically important is hydrogen (H_2). For reducing one mole of dinitrogen in water, nitrogenase is supposed to reduce 8 protons at the same time, resulting in the formation of 2 ammonia molecules, one hydrogen, and the consumption of 16 ATP.[111]

Although cultured rhizobia possess an ammonia permease and can uptake ammonium even under microaerobic conditions, ammonia synthesized by nitrogenase is believed to be translocated to the infected cell cytosol immediately by passive diffusion.[112–116] The data strongly suggest that the ammonia uptake system in rhizobia is suppressed during the differentiation of bacteria into bacteroids.[12]

Ammonia formed by nitrogenase reaction has thus been thought to diffuse freely across the SM.[19,115] However, the pH of the symbiosome space would vary in planta because of the presence of an H^+ pumping system mediated by a membrane ATP synthase.[19] If the pH in the symbiosome space is acidic, ammonia can exist mainly as ammonium ion and it would be accumulated in the symbiosome space in high concentration. In this case an NH_4^+ transporter might operate on the SM because NH_4^+ can not diffuse through the SM. A high affinity NH_4^+ transporter (Km 65 mM for methylamine) has already been reported and cloned from *Arabidopsis*.[117] Recently Day et al.[118] found such an NH_4^+ transporter on the SM in high density using the patch-clamp method.

Almost all ammonia formed in bacteroids is transferred to the plant cell, but the isolated bacteroids still retain a weak ammonia assimilating activity.[12] The significance of the primary ammonia assimilation system in bacteroids is obscure because bacteroids can be supplied amino acids from the plant cytosol. Indeed, the glutamine synthetase (GS) and glutamate synthase (GOGAT) activities are both at very low levels in bacteroids, and furthermore, these enzymes are not essential for nitrogen fixation in endosymbiosis.[119–121] Substantial activity of glutamate dehydrogenase is found in *B. japonicum* bacteroids and possibly operates in the direction of ammonia assimilation in vivo;[122,123] however, this enzyme also does not appear to be essential for nitrogen fixation activity in endosymbiosis.[124] On the other hand, isolated *B. japonicum* bacteroids can utilize glutamate and aspartate actively to support their respiration and nitrogen fixation.[122] Proline also has been suggested to be utilized efficiently by bacteroids.[125] In *R. meliloti* bacteroids, aspartate catabolism has been shown to be essential for nitrogen fixation,[126,127] even though an exogenous supply of aspartate did not enhance the nitrogen fixation of isolated bacteroids. The presence of an active uptake system of glutamate and/or aspartate has been shown in *B. japonicum* and *R. meliloti* bacteroids.[128–130] Thus, these data strongly suggest that the primary assimilation of ammonia produced from dinitrogen is not required for bacteroids, and amino nitrogen supplied from the plant cytosol is available and probably essential as a nitrogen source for bacteroids. One criticism against the utilization of amino acids by bacteroids is that an isolated symbiosome does not show the presence of an amino acid uptake system.[130] However, SM-enclosed bacteroids isolated from soybean nodules have been shown to be able to utilize exogenous glutamate under a concentration that was expected in plant cell cytosol in vivo.[122]

Plant Nitrogen Metabolism

Ammonia or ammonium ion transported from the SM is incorporated to glutamine in the cytosol via glutamine synthetase (GS), as a first step in ammonia assimilation.[7,12] The high affinity of GS to ammonia and the high concentration in the plant cytosol are indirect evidence that GS plays a key role for the rapid transport of ammonia from the symbiosome to the plant cytosol. Glutamine is then converted to glutamate via glutamate synthase (GOGAT). Direct evidence for the operation of this pathway has been obtained through ^{13}N tracer experiments.[116]

Glutamine synthetase exists in the soybean nodule cytosol at a very high concentration, ca. 2% of the cytosolic protein.[131] A gene family encodes GS in soybean plants, and one GS gene is reported to be nodule specific.[132] The abundant expression of GS in many legume nodules indicates that this system is responsible for primary assimilation of NH_3, and also suggests GOGAT may be a rate-limiting step in nitrogen assimilation. In alfalfa nodules expression of the NADH-GOGAT gene increases coincidentally with the onset of nitrogen fixation. Maximum expression of this gene appears to be required for effective nitrogen fixation.[133] Although glutamate dehydrogenase is present at high concentration in the nodule cytosol, enzyme characteristics suggest that it is not a major route to ammonia assimilation.[134] Moreover, labeling studies support a pathway of glutamate formation through glutamine.[135–137]

Various amino acids and other nitrogenous compounds can be synthesized from glutamine. However, two types of nitrogenous compound are especially important because they are translocational nitrogenous compounds loading to xylem continuously. They are amides, such as asparagine, and ureides, such as allantoin and allantoic acid.

Many indeterminate-type nodules accumulate asparagine as the major export product. Asparagine synthesis occurs through asparagine synthetase, as indicated by detailed labeling experiments.[116,138,139] Aspartate is necessary for asparagine synthetase and this compound is produced by aspartate aminotransferase. In nodules, asparagine synthetase has been detected in the plant cytosol, and aspartate aminotransferase is localized in the plastid. Both enzymes exist in nodules at high concentration.[11,12]

Ureides are the major form of nitrogen exported from determinate nodules. The two ureides observed in determinate nodules are allantoin and allantoic acids.[140,141] The latter is formed by allantoinase from the former. The biosynthesis of these compounds is more complex than that of amides. The precursor of allantoin is uric acid, and this compound is degraded by nodule specific urate oxidase (uricase II) to form allantoin.[7,142,143] There is no question about the localization of the uricase II protein, because it is a core protein of the peroxisome and is very clearly observed in peroxisomes in uninfected cells.[144] On the other hand, there is an argument on the location of biosynthesis of purine base and uric acid. High activity of purine nucleotide biosynthesis has been observed in plastids fraction in infected cells.[7,145] Xanthine dehydrogenase that can convert a purine, xanthine, to uric acid was observed by immunocytochemistry in soluble fractions of both infected and uninfected cells in the infected zone.[146] Thus, it has been proposed that purine nucleotides are formed in plastids in infected cells and transferred to uninfected cells.[7] However, plastids are also well developed in the uninfected cells in the infected zone,[12] and substantial activities of enzymes involved in de novo purine biosynthesis were found in the uninfected cell fraction, although the contamination by infected cells could not be excluded completely.[145] More experiments are necessary to determine the location of purine biosynthesis.

GENE EXPRESSION SPECIFIC IN THE SYMBIOTIC TISSUES AND CORRESPONDING METABOLIC REGULATIONS

Bacterial Genes That Affect Nitrogen Fixation

The most prominent bacterial genes in the symbiotic nitrogen fixation system are the nitrogen fixation genes (*nif* and *fix* genes), and readers should refer to recent reviews on the *nif-fix* genes of *Rhizobium* and *Bradyrhizobium*.[147,148] Here we focus on bacterial genes involved in carbon and nitrogen metabolism in relation to symbiotic performance. Such genes have primarily been identified by using natural or transposon-mediated bacterial mutants. Since the most distinctive metabolic feature of bacteroids is that their respiration is almost fully dependent on C_4-dicarboxylates supplied from host cells, it is with a good reason that the first well-defined example of bacterial mutants with altered symbiotic performance was a C4-dicarboxylate transport (*dct*) mutant. Such mutants were found to form Fix^- nodules on their corresponding host plants,[91,149,150] indicating that dicarboxylate transport is vital for symbiotic nitrogen fixation.

The *dct* genes have been intensively studied in *R. meliloti* and *R. leguminosarum.* The *dct* gene cluster is composed of three contiguous genes, *dctA, dctB* and *dctD.*[151–153] The *dctB* and *dctD* genes are proposed to encode a two-component sensor regulator system[154,155] and *dctA* encodes the dicarboxylate permease.[156] The regulation mechanism of the *dct* gene expression has been well-studied in free-living rhizobia and a detailed model for the regulation cascade leading to *dctA* gene activation has been proposed.[156] The DctB protein is assumed to be a sensor of C_4-dicarboxylate and subsequently activates (probably by phosphorylation) the DNA-binding DctD protein. The DctD protein in turn activates transcription of the *dctA* gene. This last step also requires the sigma factor, σ^{54} (a product of *rpoN*). Therefore, mutations in either the *dctA, dctB, dctD,* or *rpoN* genes results in the loss of dicarboxylate uptake.

Although *dctA* gene activation in free-living bacteria requires *dctB* and *dctD* genes, *dctA* gene expression in bacteroids can occur in mutants defective in either *dctB* or *dctD.*[153] Moreover, mutants lacking *dctB* or *dctD* function can form Fix^+ nodules, and isolated bacteroids retain the ability to transport dicarboxylates, although they exhibit reduced activity of nitrogen fixation. These results led to the suggestion that a symbiosis-specific activator(s) of *dctA* induction is present in nodules.[154] The expression of the *dctA* gene in bacteroids is independent of nitrogen fixation, and the *nifA* gene product (a regulator for *nif-fix* genes) does not appear to be involved in *dctA* gene activation.[91] Thus, the mechanism regulating *dctA* expression in the symbiotic state remains to be elucidated.

Another important question is how bacteroids oxidize dicarboxylates under low ambient oxygen concentrations in nodules. As described previously, carbon metabolism in bacteroids is an aerobic process involving the TCA cycle and ATP generation via oxidative phosphorylation. In order to adapt to the low oxygen level in nodules, the bacteroids induce a series of new electron transport proteins, including a symbiosis-specific terminal oxidase.[157] This oxidase is likely encoded by the *fix NOQP*, an operon encoding membrane-bound, cytochrome c-containing heme/copper oxidase.[158] Mutations in the *fix NOQP* gene cluster resulted in defective bacteroid development and symbiotic nitrogen fixation, suggesting this oxidase complex is specifically required for the bacteroid respiration in endosymbiosis. On the other hand, mutants defective in the *cox MNOP*, encoding an alternative cytochrome c oxidase, were found to form nitrogen-fixing nodules, and therefore this oxidase does not appear to be relevant to the symbiosis.[159]

The bacteroids of *B. japonicum* are able to utilize aspartate and glutamate through the TCA cycle for their respiration.[122] These compounds significantly enhance the nitrogen-fixing activity of bacteroids isolated from soybean nodules, and therefore may supply energy for nitrogen fixation, in planta. *B. japonicum* bacteroids were demonstrated to possess an active uptake system for glutamate.[123,130] Possibly the dicarboxylate transport system of *R. meliloti* could act as an aspartate transporter.[129]

An aspartate aminotransferase mutant (*aatA* gene) of *R. meliloti,* which is defective in aspartate catabolism, was shown to form Fix^- nodules,[126,127] suggesting that aspartate catabolism is essential for nitrogen fixation of bacteroids. Mutations in the *R. meliloti aatB* and *tatA* genes, which encode for aspartate aminotransferases distinct from *aatA*, did not affect symbiotic nitrogen fixation.[128] Therefore, *aatA* is believed to be essential for catabolism of aspartate supplied from the plant cytosol. Possible interrelationships between amino transfer in bacteroids and C_4-dicarboxylate utilization has been proposed for indeterminate nodules.[160] The malate-aspartate shuttle between the bacteroids and the plant cytosol might be relevant in indeterminate nodules. However, this shuttle does not appear to operate in determinate nodules. Detailed tracer experiments using *B. japonicum* bacteroids demonstrated that catabolism of glutamate involves amino transfer to form aspartate as an essential step, and aspartate is rapidly deaminated by aspartase.[125] Therefore, amino acid metabolism in bacteroids may differ between *Rhizobium* and *Bradyrhizobium.*

Nodulin Genes and Metabolic Regulation

Nodulin genes are defined as plant genes that are exclusively induced in root nodules.[161] However, a few genes are classified as nodulin genes because of their enhanced expression in nodules even though they are also expressed in stems and flowers[162] or in stems and uninfected roots.[163] Therefore, a better operational definition of nodulins are those genes whose expression is significantly enhanced in nodules.

Nodulins are classified into early and late nodulins according to the time of appearance of their transcripts during the nodule developmental process.[164] Early nodulins are expressed during early stages of nodule formation. Although an exact function for the various early nodulins has yet to be defined, they are thought to be involved in the infection process and/or nodule organogenesis.[165] Their expression is temporally and spatially associated to specific stages of nodule development. Some of the early nodulins are known to be induced solely by *Rhizobium* nodulation signals (Nod factors).[2,166] Alterations of hormonal balance that must be involved in the program of nodule formation also induce the expression of some early nodulins.[167] At present, however, there is no knowledge available on the involvement of these early nodulins in the metabolic differentiation specific to nodules.

Late nodulins, which are induced after the completion of nodule organogenesis and just before the onset of nitrogen fixation, are considered to play essential roles in symbiotic function (i.e., nitrogen fixation and related processes). Late nodulins have been subdivided into structural and metabolic nodulins.[168] The former includes a number of nodulins assigned as symbiosome membrane proteins, and the latter includes some nodulins that have well-defined functions, such as leghemoglobins, uricase II (nodulin 35), sucrose synthase (nodulin 100), and glutamine synthetase.

The SM has a critical importance in metabolite exchange between the host plant cell and microsymbiont. Possible relationships between the induction of SM nodulins and SM organogenesis have been proposed.[33,34] The most intensively studied SM nodulin is nodulin 26, which was isolated from soybean nodules.[35,169] Nodulin 26 is localized in the SM, but its transcription is independent from SM organogenesis. This is in contrast to another well-studied SM nodulin, nodulin 24, which appears to be involved in SM organogenesis.[35] Nodulin 26 protein has significant homology to an *E. coli* cytoplasmic membrane protein that is responsible for glycerol transport,[170] and to a tonoplast intrinsic protein (γ-TIP) found in many plant species that may function as a water channel.[171] Therefore, nodulin 26 probably is a channel protein that transports a small molecule(s) across the SM. One proposal is that nodulin 26 might be responsible for dicarboxylic acid transport across the SM, based on the observation that phosphorylation of nodulin 26 stimulates malate uptake across the SM.[172] A precise model for the assembly of the nodulin 26 protein into the SM has been proposed.[173] This model suggests that nodulin 26 is a transmembrane protein exposed to both the plant cell cytoplasm and symbiosome space and is consistent with the proposed channel function of nodulin 26.

A recently identified soybean nodulin gene, *GmN70,* was also postulated to encode an SM nodulin with a transport function.[163] Although the intracellular location of *GmN70* protein remains to be determined, *GmN70* transcripts were localized exclusively to the infected cells (Kouchi, unpublished result). The predicted amino acid sequence for *GmN70* has a typical transmembrane structure and exhibits significant homology to H^+/sulfate cotransporters recently isolated from a tropical forage legume *Stylosanthes hamata.*[173] Sequence comparison with sulfate transporters from many eukaryotes srongly suggests that *GmN70* protein is a member of a family of sulfate transporter. *GmN70* should be classified as an early nodulin gene, rather than a late nodulin, since *GmN70* transcripts appear much earlier than the onset of nitrogen fixation.[163]

Among the metabolic nodulins, the most well-defined proteins are those encoded by leghemoglobin genes. Since the literature on leghemoglobin is vast, see a recent review, reference 168, for further information. The nodule-specific uricase (uricase II) is a homotetramer of nodulin 35, which is a 33kd subunit of uricase II and the second most abundant nodulin, next to leghemoglobins, in soybean nodules.[174] Uricase (urate oxidase, EC 1.7.3.3) is an enzyme catalyzing the oxidation of uric acid to allantoin, the final step of the ureide biosynthetic pathway.

The uricase activity in soybean nodules is totally distinct from urate degradation by the diamine oxidase/peroxidase system found in uninoculated roots and hypocotyls.[175,176] Immunohistochemical studies demonstrated the location of uricase II in the peroxisomes of uninfected cells of soybean nodules.[145] The physiological significance of uricase II in determinate nodules appears clear, since the major forms for exporting reduced nitrogen out of determinate nodules are allantoin and allantoic acids. Indeed, reduction of the expression of nodulin 35 in transgenic root nodules resulted in severe nitrogen deficiency, suggesting that the reduction of ureide biosynthesis limits the availability of symbiotically fixed nitrogen to the plant.[177] Although little is known about the regulation of uricase II expression, it has been suggested that the uricase II gene is regulated by a unique mechanism distinct from that for the infected cell-specific nodulins, such as leghemoglobins and SM nodulins.[178]

Recently, uricase II activity was found to be transiently present in germinating soybean cotyledons.[179] Furthermore, nodulin 35 mRNA accumulated in cotyledons during seed development and germination. Localization of uricase II protein in the peroxisomes of germinating soybean cotyledons was also demonstrated. The sequence of the mRNA cloned from soybean cotyledons was identical to that of nodulin 35.[180] In addition, the induction of uricase II has been observed in soybean callus tissue under microaerobic conditions.[181,182] These results indicate that the expression of this gene is not solely regulated by conditions specific to nodules.

Nodulin 35 has been conventionally included with the late nodulins.[164,168] However, it should be classified in the early nodulins, because the first appearance of nodulin 35 mRNA is considerably before the onset of nitrogen fixation, although the level of the transcripts increases greatly after the onset of nitrogen fixation.[183] Thus the induction of the nodulin 35 gene appears to occur in two phases. The first phase of induction occurs at very early stages of nodule organogenesis, and the second phase, a dramatic increase of the level of the transcripts, occurs coincidentally with the onset of nitrogen fixation. The latter phase may be related to the presence of ammonia produced by bacteroids, because the level of the nodulin 35 transcripts in Fix^- soybean nodules has been shown to be significantly lower than Fix^+ nodules. On the other hand, we have recently demonstrated that nodulin 35 expression in *G. soja* roots can be induced by a purified Nod factor from *B. japonicum* (unpublished results). This result, together with the detection of a low but significant activity of uricase II in indeterminate nodules (Tajima, unpublished), suggests that uricase II plays an additional role in nodule organogenesis and/or symbiotic functioning, other than the formation of ureides as translocation products. One hypothesis for an additional function for uricase II is purine catabolism in connection with the mobilization of RNA, as suggested in the case of multifunctional peroxisomes in senescent cotyledons.[184] Constitutive expression of antisense nodulin 35 mRNA in transgenic *Vigna* roots retarded the growth of nodule peroxisomes,[177] suggesting a close relationship between organogenesis of nodule-type peroxisome and induction of uricase II.

The cDNAs for nodulin 35 have been cloned from *G. max,*[176] *Vigna aconitifolia,*[177] and *Canavalia lineata.*[185] They exhibit 85–90% identity at the nucleotide level, indicating that the nodulin 35 gene is highly conserved in legumes forming determinate nodules. In *G. max,* nodulin 35 has been suggested to comprise a small gene family[176] and two distinct genomic clones of nodulin 35 were recently isolated (Tajima, unpublished). These two genes are both expressed in nodules and cotyledons, but their expression appears to be differentially regulated during the nodule developmental process.

A de novo purine biosynthesis pathway that supplies substrates for ureide production is possibly located in infected cells of nodules. Soybean and mothbean cDNA clones encoding the first enzyme of this pathway, glutamine phosphoribosylpyrophosphate amidotransferase (PRAT), have been recently isolated from nodule cDNA libraries.[186] Although PRAT mRNA is not detected in uninfected root tissues, an exogenous supply of glutamine induces PRAT transcripts in the roots, and therefore, the accumulation of glutamine in infected cells may be involved in the induction of the PRAT gene.

Glutamine synthetase (GS) has also been intensively investigated in both determinate and indeterminate nodules. Induction of GS in nodules is developmentally regulated, but nodule-specific forms of GS have only been reported in *G. max*[132] and in *Medicago sativa.*[187] In *Pisum sativum,* several GS genes have been identified in nodules and their expression is differentially regulated in comparison to other plant organs.[188] However, nodule-specific GS gene has not been detected in *P. sativum.* An NADH-dependent glutamate synthase (NADH-GOGAT) gene has been cloned from alfalfa nodules and shown to be regulated developmentally in nodules.[133] Although this gene expression is not specific in nodules, a dramatic increase in the level of transcripts of the NADH-GOGAT gene occurs coincidentally with the onset of nitrogen fixation and is absent in either plant-determined or bacteria-determined ineffective nodules. These data suggest that the presence of ammonia secreted from bacteroids is required for the maximum expression of this gene. However, initial increase in the NADH-GOGAT expression is observed at early stages of nodule formation prior to the onset of nitrogen fixation, indicating that some signal in early nodule development might be involved in the enhanced induction of the NADH-GOGAT gene in nodules.

Nodule-specific sucrose synthase (nodulin 100) is the only known example of a nodulin with a defined function relating to carbon metabolism in nodules.[189] Nodulin 100 is an abundant protein in nodules, reflecting the large amount of sucrose that is translocated into nodules to support bacteroid respiration and to serve as carbon skeletons for the assimilation of fixed nitrogen. Recent results indicate that nodulin 100 is localized predominantly in the cytosol of the uninfected cells and inner cortical cells of soybean nodules.[190] This result, together with previous studies on the cellular distribution of enzymes of carbon metabolism in nodule tissues[65,84] shows the importance of uninfected nodule cortical cells in the breakdown of carbohydrates. Xue et al.[182] has reported that nodulin 100 is induced in soybean callus grown under low oxygen, similarly to the expression of nodulin 35.

A soybean late nodulin, GmN56, was recently identified and postulated to function in nodule carbon metabolism.[191] GmN56 shows the typical expression pattern of a late nodulin gene with mRNA localized exclusively in the infected cells of mature nodules. The predicted amino acid sequence for the GmN56 protein exhibits significant homology to bacterial isopropylmalate synthase (LeuA) and NifV protein (putative homocitrate synthase) of several nitrogen-fixing microorganisms. These enzymes have in common the ability to catalyze reactions involving acetyl-CoA and α-keto acid as substrates. The conserved domains in several LeuA and NifV proteins, which are putatively responsible to bind and/or condense these substrates, are also well conserved in the GmN56 protein. Based on these observations, it was hypothesized that GmN56 has a function related to dicarboxylate metabolism in the infected cells.

The interactions between host plant cells and microsymbionts that lead to the expression of the various nodulin genes are totally unknown. At present, a few early nodulin genes have been shown to be induced by Nod factors.[2,192,193] Nodules or nodulelike structures can be induced by purified Nod factors or by treating with phytohormone-related compounds. However, the expression of late and metabolic nodulins has not yet been reported in these structures. No leghemoglobin gene expression was detected in nodules that form spontaneously on alfalfa roots, despite the fact that these structures exhibit tissue differentiation that closely resembles genuine nodules.[194] From recent evidence it is possible that, at least in some legumes, Nod factors are sufficient to trigger the plant-encoded program of nodule organogenesis. However, to induce most of the nodulins essential for functional symbiosis, an additional series of signal exchange between plant cells and differentiating bacteroids must be required.[195] In this regard, it should be noteworthy that the high C/N ratio (high level of carbohydrate supply and nitrogen starvation) in cells might be prerequisite for the induction of nodulin genes.[196]

FUTURE OUTLOOK

Nitrogen fixing bacteroids use C_4-dicarboxylic acids as fuel and generate ATP under microaerobic conditions. Leghemoglobin and a specialized electron transport system in bacteroids allow for aerobic metabolism in the low ambient oxygen levels found in nodules. A clear focus of research has been the unique metabolism and compartmentation found in nodule cells and bacteroids. Future research will likely focus more intensively on the regulation of these systems, in particular on how these systems are induced during the nodule developmental process.

Molecular characterization of the symbiosome membrane, the gas diffusion barrier, the unloading and loading system of vascular bundles, and the function of cell organelle (e.g., mitochondria and plastids) are all important research topics that need further investigation.

Clearly an intensive area of research for the coming decade is to elucidate the various signals and signal transduction pathways operating during nodule organogenesis. The work will be on such recent major advances as the elucidation of *nod* gene functions and identification of the lipo-chitin nodulation signals. However, much work remains in identifying signal pathways operating in late stages of nodule development and function. Increased efforts to connect biochemical and genetic analysis are especially needed.

ACKNOWLEDGMENTS

We sincerely thank Drs. Gary Stacey of The University of Tennessee, Knoxville, USA, and Robert W. Ridge of International Christian University, Mitaka, Japan, for their critical reading of the manuscript.

REFERENCES

1. Dénarié, J., F. Debellé, and C. Rosenberg. 1992. Signaling and host range variation in nodulation. *Annu. Rev. Microbiol.* **46:**497–531.
2. Vijn, I., L. D. Neves, A. Van Kammen, H. J. Franssen, and T. Bisseling. 1993. Nod factors and nodulation in plants. *Science* **260:**1764–1765.
3. Truchet, G., P. Roche, P. Lerouge, J. Vasse, S. Camut, F. deBilly, J. C. Promé, and J. Dénarié. 1991. Sulphated lipo-oligosaccharide signals of *Rhizobium meliloti* elicit root nodule organogenesis in alfalfa. *Nature* **351:**670–673.
4. Stokkermans, T. J. W. and N. K. Peters. 1994. *Bradyrhizobium elkanii* lipo-oligosaccharide signals induce complete nodule structures on *Glycine soja* Siebold et Zucc. *Planta* **193:**413–420.
5. Rawsthorne, S., F. R. Minchin, R. J. Summerfield, C. Cockson, and J. Coombs. 1980. Carbon and nitrogen metabolism in legume root nodules. *Phytochemistry* **19:**341–355.
6. Stowers, M. D. 1985. Carbon metabolism in *Rhizobium* species. *Annu. Rev. Microbiol.* **39:**89–108.
7. Schubert, K. R. 1986. Products of biological nitrogen fixation in higher plants:Synthesis, transport, and metabolism. *Annu. Rev. Plant Physiol.* **37:**539–574.
8. Meller, R. B. and D. Werner. 1990. 5. Legume nodule biochemistry and function. In *Molecular Biology of Symbiotic Nitrogen Fixation,* ed. P. M. Gresshoff, pp. 111–129. CRC Press, Boca Raton, Florida.

9. Werner, D. 1991. 3. The *Rhizobium/Bradyrhizobium*-Fabales symbiosis. In *Symbiosis of Plants and Microbes,* ed. D. Werner, pp. 49–167. Chapman & Hall, London.
10. Day, D. A. and L. Copeland. 1991. Carbon metabolism and compartmentation in nitrogen-fixing legume nodules. *Plant Physiol. Biochem.* **29:**185–201.
11. Vance, C. P. and G. H. Heichel. 1991. Carbon in N_2 fixation: Limitation or equisite adaptation. *Annu. Rev. Plant Physiol. Plant Mol. Biol.* **42:**373–392.
12. Streeter, J. G. 1992. Transport and metabolism of carbon and nitrogen in legume nodules. *Adv. in Bot. Res.* **18:**130–187.
13. Werner, D. 1992. 10. Physiology of nitrogen-fixing legume nodules: Compartments and functions. In *Biological Nitrogen Fixation,* ed. G. Stacey et al., pp. 399–431. Chapman & Hall, New York.
14. Hunt, S. and D. B. Layzell. 1993. Gas exchange of legume nodules and the regulation of nitrogenase activity. *Annu. Rev. Plant Physiol. Plant Mol. Biol.* **44:**483–511.
15. Peoples, M. B., J. S. Pate, C. A. Atkins, and F. J. Bergersen. 1986. Nitrogen nutrition and xylem sap composition of peanut (*Arachis hypogaea* L. cv Virginia Bunch). *Plant Physiol.* **82:**946–951.
16. Sprent, J. I. 1980. Root nodule anatomy, type of export product and evolutionary origin in some Leguminosae. *Plant Cell and Environ.* **3:**35–43.
17. deFaria, S. M., G. P. Lewis, J. I. Sprent, and J. M Sutherland. 1989. Occurrence of nodulation in the Leguminosae. *New Phytol.* **111:**607–619.
18. Roth, E. L., K. Jeon, and G. Stacey. 1988. Homology in endosymbiotic systems: The term symbiosome. In *Molecular Genetics of Plant–Microbe Interactions,* ed. R. Palacios and D. P. S. Verma, pp. 220–225. APS Press, St. Paul, Minnesota.
19. Day, D. A. and M. K. Udvardi. 1992. Metabolite exchange across symbiosome membranes. *Symbiosis* **14:**175–189.
20. Werner, D., R. B. Mellor, M. G. Hahn, and H. Greisbach. 1985. Soybean root response to symbiotic infection. Glyceollin I accumulation in an ineffective type of soybean nodules with an early loss of the peribacteroid membrane. *Z. Naturforsch.* **40:**179–181.
21. Roth, E. L. and G. Stacey. 1989. Cytoplasmic membrane systems involved in bacterium release into soybean nodule cells as studied with two *Bradyrhizobium japonicum* mutant strains. *Eur. J. Cell Biol.* **49:**24–32.
22. Udvardi, M. K., G. D. Price, P. M. Gresshoff, and D. A. Day. 1988. A dicarboxylate transporter on the peribacteroid membrane of soybean nodules. *FEBS Lett.* 231:36–40.
23. Udvardi, M. K., L. Ou Yang, S. Young, and D. A. Day. 1990. Sugar and amino acid transport across symbiotic membranes of soybean nodules. *Mol. Plant-Microbe Interact.* **3:**334–340.
24. Ou Yang, L. J., J. Whelan, C. D. Weaver, D. M. Roberts, and D. A. Day. 1991. Protein phosphorylation stimulates the rate of malate uptake across the peribacteroid membrane of soybean nodules. *FEBS Lett.* **293:**188–190.
25. Bassarab, S. and D. Werner. 1987. Ca-2-+-dependent protein kinase activity in the peribacteroid membrane from soybean root nodules. *J. Plant Physiol.* **130:**233–241.

26. Roberts D. M. 1989. Detection of a calcium-activated protein kinase in Mougeotia by using synthetic peptide substrates. *Plant Physiol.* **91:**1613–1619.
27. Blumward, E., M. C. Fortic, P. A. Rea, D. P. S. Verma, and R. J. Poole. 1985. Presence of host-plasma membrane type H^+-ATPase in the membrane envelope enclosing the bacteroids in soybean root nodules. *Plant Physiol.* **78:**665–672.
28. Herrada, G., A. Puppo, and J. Rigaud. 1989. Uptake of metabolites by bacteroid-containing vesicles and by free bacteroids from French bean nodules. *J. of Gen. Microbiol.* **135:**3165–3171.
29. Smith. S. E. and F. A. Smith. 1990. Structure and function of the interfaces in biotrophic symbiosis as they relate to nutrient transport. *New Phytol.* **114:**1–38.
30. Kannenberg, E. J. and N. J. Brewin. 1989. Expression of a cell surface antigen is regulated by oxygen and pH of *Rhizobium leguminosarum* strain 3841. *J. Bacteriol.* **171:**4543–4548.
31. Udvardi, M. K., D. L. Lister, and D. A. Day. 1991. ATPase activity and anion transport across the peribacteroid membrane of isolated soybean symbiosomes. *Arch. Microbiol.* **156:**362–366.
32. Haaker, H., M. M. Szafran, H. J. Wassink, and M. A. Apples. 1995. Malate, aspartate and proton exchange between *Rhizobium leguminosarum* symbiosomes and its symbiotic partner *Pisum sativum.* In *Nitrogen Fixation: Fundamentals and Applications,* ed. I. A. Tikhonovich et al., pp. 565–572. Kluwer Academic Publishers, Netherlands.
33. Fortin, M. G., N. A. Morrison, and D. P. S. Verma. 1987. Nodulin 26, a peribacteroid membrane nodulin is expressed independently of the development of the peribacteroid compartment. *Nucl. Acid Res.* **15:**813–824.
34. Mellor, R. B., C. Garbers, and D. Werner. 1989. Peribacteroid membrane nodulin gene induction by *Bradyrhizobium japonicum* mutants. *Plant Mol. Biol.* **12:** 307–315.
35. Miao, G. H. and D. P. S. Verma. 1993. Soybean nodulin-26 gene encoding a channel protein is expressed only in the infected cells of nodules and is regulated differently in roots of homologous and heterologous plants. *Plant Cell* **5:**781–794.
36. Werner, D., A. Jacobi, P. Müller and T. Winzer, 1995. Symbiosomes under control of plant and microsymbiont genes. In *Nitrogen Fixation: Fundamentals and Applications,* ed. I. A. Tikhonovich et al., pp. 449–453. Kluwer Academic Publishers, Netherlands.
37. Puppo, A., G. Herrada, and J. Rigaud. 1991. Lipid peroxidation in peribacteroid membranes from french-bean nodules. *Plant Physiol.* **96:**826–830.
38. Jacobi, A., P. Katinakis, and D. Werner. 1994. Artificially induced senescence of soybean root nodules affects different polypeptides and nodulins in the symbiosome membrane compared to physiological ageing. *J. Plant Physiol.* **144:**533–540.
39. Ozawa, T. and T. Tsujii. 1993. A possible role for polyamines in the repression of growth of *Bradyrhizobium japonicum* bacteroids in soybean nodules. *Plant Cell Physiol.* **34:**899–904.
40. Ozawa, T. and T. Tsujii. 1994. Isolation of a *Bradyrhizobium japonicum* growth inhibitor from soybean roots. *Jpn. J. Soil Sci. Plant Nutr.* **65:**654–659.

41. Pankhurst, C. E., W. T. Jones, and A. S. Craig. 1982. Bactericidal effect of *Lotus pedunculatus* root flavolan on fast-growing lotus rhizobia. *J. Gen. Microbiol.* **128:**1567–1576.
42. Tjepkema, J. D. and C. S. Yocum. 1974. Measurement of oxygen partial pressure within soybean nodules by oxygen microelectrodes. *Planta* **119:**351–360.
43. Witty, J. F., L. Skot, and N. P. Revsbech. 1987. Direct evidence for changes in the resistance of legume root nodules to O2 diffusion. *J. Exp. Bot.* **38:**1129–1140.
44. Layzell, D. B. and S. Hunt. 1990. Oxygen and the regulation of nitrogen fixation in legume nodules. *Physiol. Plant* **80:**322–327.
45. Kuzma, M. M. and D. B. Layzell. 1994. Acclimation of soybean nodules to change in temperature. *Plant Physiol.* **106:**263–270.
46. Bergersen, F. J. and G. L. Turner. 1980. Properties of terminal oxidase systems of bacteroids from root nodules of soybean and cowpea and of N_2-fixing bacteria grown in continuous culture. *J. Gen. Microbiol.,* **118:**235–252.
47. Rawsthorne, S. and T. A. LaRue. 1986. Preparation and properties of mitochondria from cowpea nodules. *Plant Physiol.* **81:**1092–1096.
48. Bergersen, F. J. 1962. Oxygenation of leghemoglobin in soybean root-nodules in relation to the external oxygen tension. *Nature* **194:**1059–1061.
49. Appleby, C. A. 1984. Leghemoglobin and *Rhizobium* respiration. *Annu. Rev. Plant Physiol.* **35:**443–478.
50. Layzell, D. B., S. Hunt, and G. R. Palmer. 1990. Mechanism of nitrogenase inhibition in soybean nodules. Pulse-modulated spectroscopy indicates that nitrogenase activity is limited by O_2. *Plant Physiol.* **92:**1101–1107.
51. Becana, M. and R. V. Klucas. 1992. Oxidation and reduction of leghemoglobin in root nodules of leguminous plants. *Plant Physiol.* **98:**1217–1221.
52. Saari, L.L. and R.V. Klucas. 1984. Ferric leghemoglobin reductase from soybean nodules. *Arch. Biochem. Biophys.* **231:**102–113.
53. Ji, L., M. Becana, G. Sarath, and R. V. Klucas. 1994. Cloning and sequence analysis of a cDNA encoding ferric leghemoglobin reductase from soybean nodules. *Plant Physiol.* **104:**453–459.
54. Kanayama, Y. and Y. Yamamoto. 1990. Inhibition of nitrogen fixation in soybean plants supplied with nitrate. II. Accumulation and properties of nitrosylleghemoglobin in nodules. *Plant Cell Physiol.* **31:**207–214.
55. Pankhurst, C. E. and Sprent, J. L. 1975 . Surface features of soybean root nodules. *Protoplasma* **85:**85–98.
56. Castillo, L. D. and D. B. Layzell, 1994. Drought stress, permeability to O_2 diffusion and the respiratory kinetics of soybean root nodules. *Plant Physiol.* **106:**949–955.
57. Tajima, S. and K. Kouzai. 1989. Nucleotide pools in soybean nodule tissues, a survey of NAD(P)/NAD(P)H ratios and energy charge. *Plant Cell Physiol.* **30:**589–593.
58. Oresnik, I. J. and D. B. Layzell. 1994. Composition and distribution of adenylates in soybean (*Glycine max* L.) nodule tissue. *Plant Physiol.* **104:**217–225.

59. de Lima, M. L., I. J. Oresnik, S. M. Fernando, S. Hunt, R. Smith, D. H. Turpin, and D. B. Layzell. 1994. The relationship between nodule adenylates and the regulation of nitrogenase activity by O_2 in soybean. *Physiologia Plantarum* **91:**687-695.

60. Weisz, P. R. and T. R. Sinclair. 1988. A rapid nondestructive assay to quantify soybean nodule gas permeability. *Plant Soil.* **105:**69–78.

61. Denison, R. F. and D. B. Layzell. 1991. Measurement of legume nodule respiration and O_2 permeability by noninvasive spectrophotometry of leghemoglobin. *Plant Physiol.* **96:**137–143.

62. Kuzma, M. M., S. Hunt, and D.B. Layzell. 1993. The role of O_2 in the limitation and inhibition of nitrogenase activity and respiration rate in individual soybean nodules. *Plant Physiol.* **101:**161–169.

63. Castillo, D. L., S. Hunt, and D. B. Layzell. 1992. O_2 regulation and O_2 limitation of nitrogenase activity in root nodules of pea and lupin. *Physiol. Plant* **86:**269–278.

64. Kouchi, H. and T. Yoneyama. 1986. Metabolism of ^{13}C-labelled photosynthate in plant cytosol and bacteroids of root nodules of *Glycine max. Physiol. Plant* **68:** 238–244.

65. Kouchi, H. and T. Higuchi. 1988. Carbon flow from nodulated roots to the shoots of soybean (*Glycine max* L. Merr.) plants: An estimation of the contribution of current photosynthate to ureides in the xylem stream. *J. Exp. Bot.* **39:**1015–1023.

66. Vance, P. C., G. H. Heichel, J. W. Bryan, and L. E. Johnson. 1979. Nitrogen fixation, nodule development, and vegitative regrowth of alfalfa (*Medicago sativa* L.) following harvest. *Plant Physiol.* **64:**1–8.

67. Hartwig, U., B. C. Boller, B. Bauer-Hoch, and J. Nosberger. 1990. The influence of carbohydrate reserves on the response of nodulated white clover to defoliation. *Ann. Bot.* **65:**721–725.

68. Finn, G. A. and W. A. Brun. 1982. Effect of atomospheric CO_2 enrichment on growth, nonstructural carbohydrate content, and root nodule activity in soybean. *Plant Physiol.* **69:**327–331.

69. Drevon, J. J., J. P. Gaudillere, J. P. Bernoud, F. Jardinet, and M. Evrard. 1991. Influence of photon flux density and fluctuation on the nitrogen fixing *Glycine max* (L. Merr)-*Bradyrhizobium japonicum* symbiosis in a controlled environment. *Agronomie* **11:**193–199.

70. Gordon, A. J., W. Kessler, and F. R. Minchin. 1990. Defoliation-induced stress in nodules of white clover. I. Changes in physiological parameters and protein synthesis. *J. Exp. Bot.* **41:**1245–1253.

71. Sung, L., A. H. M. Moloney, S. Hunt, and D. B. Layzell. 1991. The effect of excision on O_2 diffusion and metabolism in soybean nodules. *Physiol. Plant* **83:**67–74.

72. Riesmeier, J. W., L.Willmitzer, and W. B. Frommer. 1994. Evidence for an essential role of the sucrose transporter in phloem loading and assimilate partitioning. *EMBO J.* **13:**1–7.

73. Selker, J. M. L. and E. H. Newcomb. 1985. Spatial relationships between uninfected and infected cells in root nodules of soybean. *Planta* **165:**446–454.

74. Selker, J. M. 1988. Three-dimensional organization of uninfected tissue in soybean root nodules and its relation to cell specialization in the central region. *Protoplasma* **147:**178–190.

75. Kouchi, H., M. Tsukamoto, and S. Tajima. 1989. Differential expression of nodule specific (nodulin) genes in the infected, uninfected and cortical cells of soybean (*Glycine max*) root nodules. *J. Plant Physiol.* **135:**608–617.

76. Zammit, A. and L. Copeland. 1993. Immunocyctochemical localisation of nodule-specific sucrose synthase in soybean nodules. *Aust. J. Plant Physiol.* **20:**25–32.

77. Vella, J. and L. Copeland. 1990. UDP-glucose pyrophosphorylase from the plant fraction of nitrogen-fixing soybean nodules. *Physiol. Plant* **78:**140–146.

78. Peterson, J. B. and H. J. Evans. 1979. Phosphoenol-pyruvate carboxylase from soybean nodule cytosol. Evidence for isoenzymes and kinetics of the most active component. *Biochim. Biophys. Acta* **567:**445–452.

79. Vance, C. P., K. L. M. Boylan, C. A. Maxwell,G. H. Heichel, and L. L. Hardman. 1985. Transport and partitioning of CO_2 fixed by root nodules of ureide and amide producing legumes. *Plant Physiol.* **78:**774–778.

80. Tajima, S. and T. A. LaRue. 1982. Enzymes for acetaldehyde and ethanol formation in legume nodules. *Plant Physiol.* **70:**388–392.

81. Davies, D. D. 1990. Anaerobic metabolism and the production of organic acids. In *The Biochemistry of Plants, Vol 2. Metabolism and Respiration,* ed. D. D. Davies, pp. 581–611, Academic Press, New York.

82. Copeland, L., J. Vella, and Z. Hong. 1989. Enzymes of carbohydrate metabolism in soybean nodules. *Phytochemistry* **28:**57–61.

83. Suganuma, N., M. Kitou, and Y. Yamamoto. 1987. Carbon metabolism in relation to cellular organization of soybean root nodules and respiration of mitochondria aided by leghemoglobin. *Plant Cell Physiol.* **28:**113–122.

84. Kouchi, H., K. Fukai, H. Katagiri, K. Minamisawa, and S. Tajima. 1988. Isolation and enzymological characterization of infected and uninfected cell protoplasts from root nodules of *Glycine max. Physiol. Plant* **73:**327–334.

85. Rawsthorne, S. and T. A. LaRue. 1986. Metabolism under microaerobic conditions of mitochondria from cowpea nodules. *Plant Physiol.* **81:**1097–1102.

86. Suganuma, N. and Y. Yamamoto. 1987. Respiratory metabolism of mitochondria in soybean root nodules. *Soil Sci. and Plant Nutr.* **33:**93–101.

87. Phillips, D. V., D. O. Wilson, and D. E. Dougherty. 1984. Soluble carbohydrates in legumes and nodulated nonlegumes. *J. Agr. Food Chem.* **32:**1289–1291.

88. Streeter, J. G. 1980. Carbohydrates in soybean nodules. II. Distribution of compounds in seedlings during the onset of nitrogen fixation. *Plant Physiol.* **66:**471–476.

89. Tikhonovich, I. A., A. Y. Borisov, I. A. Chvabauskene, N. N. Kamardin, L. V. Kravchenko, V. K. Lebsky, F. Minchin, E. V. Morzina, V. I. Romanov, L. Skot, S.A. Tchetkova, and V. E. Tsiganov. 1995. Plant genetical control at the crucial steps of symbiosis. In *Nitrogen Fixation: Fundamentals and Applications,* ed. I. A. Tikhonovich et al., pp. 461–466. Kluwer Academic Publishers, Netherlands.

90. Salminen, S. O. and J. G. Streeter. 1986. Enzymes of α,α-trehalose metabolism in soybean nodules. *Plant Physiol.* **81:**538–541.

91. Jording, D., P. K. Sharma, R. Schmidt, T. Engelke, C. Uhde, and A. Pühler. 1993. Regulatory aspects of the C_4-dicarboxylate transport in *Rhizobium meliloti:* Transcriptional activation and dependence on effective symbiosis. *J. Plant Physiol.* **141:**18–27.

92. Ronson, C. W., P. Lyttleton and J. G. Robertson. 1981. C_4-dicarboxylate transport mutants of *Rhizobium trifolii* form ineffective nodules on Trifolium repens. *Proc. Natl. Acad. Sci. USA* **78:**4284–4288.

93. Hornez, J. P., M. Timinouni, C. Defives, and J. C. Derieux. 1994. Unaffected nodulation and nitrogen fixation in carbohydrate pleiotropic mutants of *Rhizobium meliloti. Current Microbiol.* **28:**225–229.

94. Tajima, S., H. Sasahara, H. Kouchi, T. Yoneyama, and J. Ishizuka. 1985. Effects of oxygen concentration and leghemoglobin on organic acid degradation by isolated soybean nodule bacteroids. *Agric. Biol. Chem.* **49:**3473–3479.

95. Tajima, S., I. Kimura, and H. Sasahara. 1986. Succinate metabolism of isolated soybean nodule bacteroids at low oxygen concentration. *Agric. Biol. Chem.* **50:**1009–1014.

96. McDermott, T. R., S. M. Griffith, C. P. Vance, and P. H. Graham. 1989. Carbon metabolim in *Bradyrhizobium japonicum* bacteroids. *FEMS Microbiol. Rev.* **63:**327–340.

97. Miller, R. W., D. G. McRae, A. Al-Jobore, and W. B. Berndt. 1988. Respiration supported nitrogenase activity of isolated *Rhizobium meliloti* bacteroids. *J. Cell Biochem.* **38:**35–49.

98. Miller, R. W., D. G. McRae, and K. Joy. 1991. Glutamate and γ-aminobutyrate metabolism in isolated *Rhizobium meliloti* bacteroids. *Mol. Plant-Microbe Interact.* **4:**37–45.

99. Fitzmaurice, A. M. and F. O'Gara. 1993. A *Rhizobium meliloti* mutant, lacking a functional γ-aminobutyrate (GABA) bypass, is defective in glutamate catabolism and symbiotic nitrogen fixation. *FEMS Microbiol. Lett.* **109:**195–202.

100. Salminen, S. O. and J. G. Streeter. 1990. Factors affecting to the accumulation of glutamate in *Bradyrhizobium japonicum* bacteroids under microaerobic conditions. *J. Gen. Microbiol.* **136:**2119–2126.

101. Kahn, M. L., M. Mortimer, K. S. Park, and W. Zhang. 1995. Carbon metabolism in the *Rhizobium*-legume symibiosis. In *Nitrogen Fixation: Fundamentals and Applications,* ed. I. A. Tikhonovich et al., pp. 525–532. Kluwer Academic Publishers, Netherlands.

102. Kimura, I. and S. Tajima. 1989. Presence and characteristics of NADP-malic enzyme in soybean nodule bacteroids. Soil Sci. *Plant Nutr.* **35:**271–279.

103. Driscoll, B. T. and T. M. Finan. 1993. NAD^+-dependent malic enzyme of *Rhizobium meliloti* is required for symbiotic nitrogen fixation. *Mol. Microbiol.* **7:**865–873.

104. Discoll, B. T., M. Osteras, and T. M. Finan. 1995. Malic enzyme of *Rhizobium meliloti.* In *Nitrogen Fixation: Fundamentals and Applications,* ed. I. A. Tikhonovich et al., pp. 539–544. Kluwer Academic Publishers, Netherlands.

105. Day, D.A., R.G. Quinnell, and F.J. Bergersen. 1994. An hypothesis on the role of malic enzyme in symbiotic nitrogen fixation in soybean nodules. In *Symbiotic Nitrogen Fixation,* ed. P. H. Graham et al., pp. 159–164. Kluwer Academic Publishers, Dordrecht.

106. Kobayashi, K., S. Doi, S. Negoro, I. Urabe, and H. Okada. 1989. Structure and properties of malic enzyme from *Bacillus searothermophilus. J. Biol. Chem.* **264:**3200–3205.

107. Bergerson, F. J. and G. L. Turner. 1975. Leghemoglobin and the supply of O_2 to nitrogen fixing root nodule bacteroids: Presence of two oxidase systems and ATP production at low free O_2 concentrations. *J. Gen. Microbiol.* **91:**345–354.

108. O'Brian, M. R. and R. J. Maier. 1989. Molecular aspects of the energetics of nitrogen fixation in *Rhizobium*-legume symbiosis. *Biochim. Biophys. Acta* **974:**229–246.

109. Streeter, J. G. and S. O. Salminen. 1993. Distribution of the two types of polysaccharide formed by *Bradyrhizobium japonicum* bacteroids in nodules of field-grown soybean plants (*Glycine max* (L.) Merr.). *Soil Biol. Biochem.* **25:**1027–1032.

110. Streeter, J. G., S. O. Salminen, J. E. Beuerlien, and W. H. Schmidt. 1994. Factors influencing the synthesis of polysaccharide by *Bradyrhizobium japonicum* bacteroids in field-grown soybean nodules. *Appl. Environ. Microbiol.* **60:** 2939–2943.

111. Phillips, D. A. 1980. Efficiency of symbiotic nitrogen fixation in legumes. *Annu. Rev. Plant Physiol.* **31:**29–49.

112. O'Hara, G. W., I. T. Riley, A. R. Glenn, and M. J. Dilworth. 1985. The ammonium permease of *Rhizobium leguminosarum* MNF 3841. *J. Gen. Microbiol.* **131:**757–764.

113. Jin, H. N., A. R. Glenn, and M. J. Dilworth. 1988. Ammonium uptake by cowpea *Rhizobium* strain MNF 2030 and *Rhizobium trifolii* MNF 1001. *Arch. Microbiol.* **149:**308–311.

114. Howitt, S. M., M. K. Udvardi, D. A. Day, and P. M. Gresshoff. 1986. Ammonia transport in free-living and symbiotic *Rhizobium* sp. ANU289. *J. Gen. Microbiol.* **132:**257–261.

115. Streeter, J. G. 1989. Estimation of ammonium concentration in the cytosol of soybean nodules. *Plant Physiol.* **90:**779–782.

116. Meeks. J. C., C. P Wolk, N. Schilling, P. W. Shaffer, Y. Avissar, and W. S. Chien. 1978. Initial organic products of fixation of ^{13}N dinitrogen by root nodules of soybean (*Glycine max*). *Plant Physiol.* **61:**980–983.

117. Ninnemann, O., J. Jauniaux, and W. B. Frommer. 1994. Identification of a high affinity NH_4^+ transporter from plants. *EMBO J.* **13:**3464–3471.

118. Day, D. A., L. Whitehead, J. H. M. Hendriks and S. D. Tyerman. 1995. Nitrogen and carbon exchange across symbiotic membranes from soybean nodules. In *Nitrogen Fixation: Fundamentals and Applications,* ed. I. A. Tikhonovich et al., pp. 557–564. Kluwer Academic Publishers, Netherlands.

119. Morrett, E., S. Moreno, and G. Epsen. 1985. Impaired nitrogen fixation and glutamine synthesis in methionine sulfoximine sensitive (MS^s) mutants of *Rhizobium phaseoli. Mol. Gen. Genetics* **200:**229–234.

120. deBruijn, F. J., S. Rossbach, M. Schneider, P. Ratet, S. Messmer, W. W. Szeto, F. M. Ausubel, and J. Schell. 1989. *Rhizobium meliloti* 1021 has three differentially regulated loci involved in glutamine biosynthesis, none of which is essential for symbiotic nitrogen fixation. *J. Bacteriol.* **171:**1673–1682.

121. Lewis, T. A., G. Rigoberto, and J. L. Botsford. 1990. *Rhizobium meliloti* glutamine synthase: Cloning and initial characterization of the *glt* locus. *J. Bacteriol.* **172:**2413–2420.

122. Kouchi, H., K. Fukai, and A. Kihara. 1991. Metabolism of aspartate and glutamate in soybean nodule bacteroids. *J. Gen. Microbiol.* **137:**2901–2910.

123. Salminen, S. O. and J. G. Streeter. 1987. Involvement of glutamate in the respiratory metabolism of *Rhizobium japonicum* bacteroids. *J. Bacteriol.* **169:**495–499.

124. Lane, M., J. Meade, S. S. Manian, and F. O'Gara. 1986. Expression and regulation of the *Escherichia coli* glutamate dehydrogenase gene (*gdh*) in *Rhizobium japonicum. Arch. Microbiol.* **144:**29–34.

125. Kohl, D. H., K. R. Schubert, M. B. Carter, C. H. Hagedorn, and G. Shearer. 1988. Proline metabolism in N_2-fixing nodules: Energy transfer and regulation of purine synthesis. *Proc. Natl. Acad. Sci. USA* **85:**2036–2040.

126. Rastogi, V. K. and R. J. Watson. 1991. Aspartate aminotransferase activity is required for aspartate catabolism and symbiotic nitrogen fixation in *Rhizobium meliloti. J. Bacteriol.* **173:**2879–2887.

127. Watson, R. J. and V. K. Rastogi. 1993. Cloning and nucleotide sequenceing of *Rhizobium meliloti* aminotransferase genes: an aspartate aminotransferase required for symbiotic nitrogen fixation is atypical. *J. Bacteriol.* **175:**1919–1928.

128. Alfano, J. R. and M. L. Kahn. 1993. Isolation and characterization of a gene coding for a novel aspartate aminotransferase from *Rhizobium meliloti. J. Bacteriol.* **175:**4186–4196.

129. Watson, R. J., Y.-K. Chan, R. Wheatcroft, A.-F. Yang, and S. Han. 1988. *Rhizobium meliloti* genes required for C_4-dicarboxylate transport and symbiotic nitrogen fixation are located on a megaplasmid. *J. Bacteriol.* **170:**927–934.

130. Udvardi, M. K., C. L. Salom, and D. A. Day. 1988. Transport of L-glutamate across the bacteroid membrane but not the peribacteroid membrane from soybean root nodules. *Mol. Plant-Microbe Interact.* **1:**250–254.

131. McParland, R. H., J. G. Guevara, R. R. Becker, and H. J. Evans. 1976. The purification and properties of the glutamine synthetase from the cytosol of soya-bean root nodules. *Biochem. J.* **153:**597–606.

132. Roche, D. S., J. Temple, and C. Sengupta-Gopalan. 1993. Two classes of differentially regulated glutamine synthetase genes are expressed in the soybean nodule: a nodule-specific class and a constitutively expressed class. *Plant Mol. Biol.* **22:**971–983.

133. Gregerson, R. G., S. S. Miller, S. N. Twary, J. S. Gantt, and C. P. Vance. 1993. Molecular characterization of NADH-dependent glutamate synthase from alfalfa nodules. *Plant Cell* **5**:215–226.

134. Stone, S. R., L. Copeland, and I. R. Kennedy. 1979. Glutamate dehydrogenase of lupin nodules: Purification and properties. *Phytochemistry* **18**:1273–1278.

135. Ohyama, T. and K. Kumazawa. 1980. Nitrogen assimilation in soybean nodules I. The role of GS/GOGAT system in the assimilation of ammonia produced by N_2-fixation. *Soil Sci. Plant Nutr.* **26:**109–115.

136. Ohyama, T. and K. Kumazawa. 1980. Nitrogen assimilation in soybean nodules II. N_2 assimilation in bacteroid and cytosol fractions of soybean nodules. *Soil Sci. Plant Nutr.* **26:**205–213.

137. Atkins, C. A., P. J. Storer, and J. S. Pate. 1988. Pathways of nitrogen assimilation in cowpea nodules studied using $^{15}N_2$ and allopurinol. *Plant Physiol.* **86:**204–207.

138. Fujihara, S. and M. Yamaguchi. 1980. Asparagine formation in soybean nodules. *Plant Physiol.* **66:**139–141.

139. Minamisawa, K., Y. Arima, and K. Kumazawa. 1986. Characteristics of asparagine pool in soybean nodules in comparison with ureide pool. *Soil Sci. Plant Nutr.* **32:**1–14.

140. Yoneyama, T., M. Karasuyama, H. Kouchi and J. Ishizuka. 1985. Occurrence of ureide accumulation in soybean plants. Effect of nitrogen fertilization and N_2 fixation. *Soil Sci. Plant Nutr.* **31:**133–140.

141. Streeter, J. G. 1979. Allantoin and allantoic acid in tissues and stem exudate from field-grown soybean plants. *Plant Physiol.* **63:**478–480.

142. Tajima, S. and Y. Yamamoto. 1975. Enzymes of purine catabolism in soybean nodules. *Plant Cell Physiol.* **16:**271–282.

143. Schubert, K. R. 1981. Enzymes of purine biosynthesis and catabolism in *Glycine max*. I. Comparison of activities with N_2 fixation and composition of xylem exudate during nodule development. *Plant Physiol.* **68:**1115–1122.

144. VandenBosch, K. A. and E. H. Newcomb. 1986. Immunogold localization of nodule-specific uricase in developing soybean root nodules. *Planta* **167:**425–436.

145. Shelp. B. J., C. A. Atkins, P. J. Storer, and D. T. Canvin. 1983. Cellular and subcellular organization of pathways of ammonia assimilation and ureide synthesis in nodules of cowpea (*Vigna unguiculata* L. Walp.). *Arch. Biochem. Biophys.* **224:**429–441.

146. Datta, D. B., E. W. Triplett, and E. H. Newcomb. 1991. Localization of xanthine dehydrogenase in cowpea root nodules: Implications for the interaction between cellular compartments during ureide biogenesis. *Proc. Natl. Acad. Sci. USA* **88:**4700–4702.

147. Barbour, W. M., S.-P. Wang, and G. Stacey. 1991. Molecular genetics of *Bradyrhizobium* symbiosis. In *Biological Nitrogen Fixation,* eds. G. Stacey et al., pp. 625–647. Chapman & Hall, New York.

148. Merrik, M. J. 1991. Regulation of nitrogen fixation genes in free-living and symbiotic bacteria. In *Biological Nitrogen Fixation,* eds. G. Stacey et al., pp. 835–876. Chapman & Hall, New York.

149. Arwas, R., I. A. McKay, F. R. P. Rowney, M. J. Dilworth, and A. R. Glenn. 1985. Properties of organic acid utilization mutants of *Rhizobium leguminosarum* strain 300. *J. Gen. Microbiol.* **131:**2059–2066.

150. Engelke, T., M. N. Jagadish and A. Puhler. 1987. Biochemical and genetical analysis of *Rhizobium meliloti* mutants defective in C_4-dicarboxylate transport. *J. Gen. Microbiol.* **133:**3019–3029.
151. Ledebur, H., B. Gu, J. Sojda, and B. T. Nixon. 1990. *Rhizobium meliloti* and *Rhizobium leguminosarum* dctD gene products bind to tandem sites in an activation sequence located upstream of s^{54}-dependent dctA promoters. *J. Bacteriol.* **172:**3888–3897.
152. Watson, R. J.. 1990. Analysis of the C-4—dicarboxylate transport genes of *Rhizobium meliloti*: nucleotide sequence and deduced products of dctA, dctB, and dctD. *Mol. Plant-Microbe Interact.* **3:**174–181.
153. Yarosh, O. K., T. C. Charles, and T. M. Finan. 1989. Analysis of C_4-dicarboxylate transport genes in *Rhizobium meliloti. Mol. Microbiol.* **3:**813–823.
154. Nixon, B. T., C. W. Ronson, and F. M. Ausubel. 1986. Two-component regulatory stystem responsive to environmental stimuli share strongly conserved domains with the nitrogen assimilation regulatory genes *ntrB* and *ntrC. Proc. Natl. Acad. Sci. USA* **83:**7850–7854.
155. Ronson, C. W., B. T. Nixon, L. M. Albright, and F. M. Ausubel. 1987. Conserved domains in bacterial regulatory proteins that respond to environmental stimuli. *Cell* **49**:579–581.
156. Ronson, C. W., P. M. Astwood, B. T. Nixon, and F. M. Ausubel. 1987. Deduced products of C_4-dicarboxylate transport regulatory genes of *Rhizobium leguminosarum* are homologous to nitrogen regulatory gene products. *Nucl. Acid Res.* **15:**7921–7934.
157. Bergerson, F. J. and G. L. Turner. 1975. Leghemoglobin and the supply of O_2 to nitrogen fixing root nodule bacteroids: Presence of two oxidase systems and ATP production at low free O_2 concentrations. *J. Gen. Microbiol.* **91:**345–354.
158. Preisig, O., D. Anthamatten, and H. Hennecke. 1993. Genes for a microaerobically induced oxidase complex in *Bradyrhizobium japonicum* are essential for a nitrogen-fixing endosymbiosis. *Proc. Natl. Acad. Sci. USA* **90:**3309–3313.
159. Bott, M., O. Preisig, and H. Hennecke. 1992. Genes for a second terminal oxidase in *Bradyrhizobium japonicum. Arch. Microbiol.* **158:**335–343.
160. Kahn, M. L., J. Kraus, and J. E. Somerville. 1985. A model of nutrient exchange in the Rhizobium-legume symbiosis. In *Nitrogen Fixation Research Progress,* ed. H. J. Evans et al., pp. 193–199. Martinus Nijhoff, Boston.
161. Van Kammen, A. 1984. Suggested nomenclature for plant genes involved in nodulation and symbiosis. *Plant Mol. Biol. Rep.* **2:**43–45.
162. Scheres, B., F. van Engelen, E. Van der Knaap, C. Van de Wiel, A. Van Kammen, and T. Bisseling. 1990. Sequential induction of nodulin gene expression in the developing pea nodule. *Plant Cell* **2:**687–700.
163. Kouchi, H. and S. Hata. 1993. Isolation and characterization of novel nodulin cDNAs representing genes expressed at early stages of soybean nodule development. *Mol. Gen. Genetics.* **238:**106–119.
164. Franssen, H. J., J.-P. Nap, and T. Bisseling. 1991. Nodulins in root nodule development. In *Biological Nitrogen Fixation,* ed. G. Stacey et al., pp. 598–624. Chapmann and Hall, New York.

165. Nap, J. P. and T. Bisseling. 1990. Developmental biology of a plant-procaryote symbiosis: The legume root nodule. *Science* **250:**948–954.

166. Crespi, M. D., E. Jurkevitch, M. Poiret, Y. Daubentoncarafa, G. Petrovics, E. Kondorosi, and A. Kondorosi. 1994. Enod40, a gene expressed during nodule organogenesis, codes for a non-translatable RNA involved in plant growth. *EMBO J.* **13:**5099–5112.

167. Hirsch, A. M., T. V. Bhuvaneswari, J. G. Torrey and T. Bisseling. 1989. Early nodulin genes are induced in alfalfa root outgrowth elicited by auxin transport inhibitors. *Proc. Natl. Acad. Sci. USA* **86:**1244–1248.

168. Delauney, A. J. and D. P. S. Verma. 1988. Cloned nodulin genes for symbiotic nitrogen fixation. *Plant Mol. Biol. Rep.* **6**:279–285.

169. Weaver, C. D., N. H. Shomer, C. F. Louis, and D. M. Roberts. 1994. Nodulin 26, a nodule-specific symbiosome membrane protein from soybean, is an ion channel. *J. Biol. Chem.* **269**:17858–17862.

170. Sweet, G., C. Gandor, R. Voegele, N.A. Wittekindt, J. Beuerle, V. Truniger, E.C.C. Lin, and B. Winfried. 1990. Glycerol facilitator of *Escherichia coli:* Cloning of glpF and identification of the glpF product. *J. Bacteriol.* **172:**424–430.

171. Johnson, K. D., H. Hofte, and M. J. Chrispeels. 1990. An intrinsic tonoplast protein of protein storage vacuoles in seeds is structurally reklated to bacterial solute transporter(GlpF). *Plant Cell* **2:**525–532.

172. Miao, G. H., Z. Hong, and D. P. S. Verma. 1992. Topology and phosphorylation of soybean nodulin-26, an intrinsic protein of the peribacteroid membrane. *J. Cell Biol.* **118:**481–490.

173. Smith, R. W., P. M. Ealing, M. J. Hawkesford, and D. T. Clarkson. 1995. Plant members of a family of sulfate transporters reveal functional subtype. *Proc. Natl. Acad. Sci. USA* **92:**9373-9377.

174. Bergmann, H., E. P. Preddie, and D. P. S. Verma. 1983. Nodulin-35: a subunit of specific uricase (uricase II) induced and localized in the uninfected cells of soybean nodules. *EMBO J.* **2:**2333–2339.

175. Tajima, S., N. Kato, and Y. Yamamoto. 1983. Cadaverine involved in urate degrading activity (uricase activity) in soybean radicles. *Plant Cell Physiol.* **24:**247–253.

176. Nguyen, T., M. Zelechowska, V. Foster, H. Bergmann, and D. P. S. Verma. 1985. Primary structure of the soybean nodulin-35 gene encoding uricase II localized in the peroxisomes of uninfected cells of nodules. *Proc. Natl. Acad. Sci. USA* **82:**5040–5044.

177. Lee, N. G., B. Stein, H. Suzuki, and D. P. S. Verma. 1993. Expression of antisense nodulin-35 RNA in *Vigna aconitifolia* transgenic root nodules retards peroxisome development and affects nitrogen availability to the plant. *Plant J.* **3:**599–606.

178. Mauro, V. P. and D. P. S. Verma. 1988. Transcriptional activation in nuclei from uninfected soybean of a set of genes involved in symbiosis with *Rhizobium. Mol. Plant-Microbe Interact.* **1:**46–51.

179. Tajima, S., H. Ito, K. Tanaka, T. Nanakado, A. Sugimoto, H. Kouchi, and K. Okazaki. 1991. Soybean cotyledons contain a uricase that cross-reacts with antibodies raised against the nodule uricase (nod-35). *Plant Cell Physiol.* **32:**1307–1311.

180. Tajima, S., K. Takane, K Seki, K. Ohkawa, K. Okazaki, and H. Kouchi. 1995. Expression of uricase gene family in soybean tissues & comparison of the uricases in various legume nodules. In *Nitrogen Fixation: Fundamentals and Applications,* ed. I. A. Tikhonovich et al., pp. 511. Kluwer Academic Publishers, Netherlands.

181. Larsen, K. and B. U. Jochimsen. 1986. Expression of nodule-specific uricase in soybean callus tissue is regulated by oxygen. *EMBO J.* **5:**15–19.

182. Xue, Z., K. Larsen, and B.U. Jochimsen. 1991. Oxygen regulation of uricase and sucrose synthase synthesis in soybean callus tissue is exerted at the mRNA level. *Plant Mol. Biol.* **16:**899–906.

183. Tajima, S., K. Tanaka, K. Takane, A. Sugimoto, K. Okazaki and H. Kouchi. 1992. Soybean nodule uricase gene (nodulin-35) is expressed in cotyledons during seed development and early germination. In *New Horizons in Nitrogen Fixation,* ed. R. Palacios rt al., pp. 373. Kluwer Academic Publishers, Dordrecht.

184. Vincentini, F. and P. Matile. 1993. Gerontosomes, a multifunctional type of peroxisome in senescent leaves. *J. Plant Physiol.* **42:**50–56.

185. Kim, H. B. and C. S. An. 1993. Nucleotide sequences and expression of cDNA encoding uricase II in *Canavalia lineata. Korean J. Bot.* **36:**415–423.

186. Kim, J. H., A. J. Delauney and D. P. S. Verma. 1995. Control of de novo purine biosynthesis genes in ureide-producing legumes: Induction of glutamine phosphoribosylpyrophosphate amidotransferase gene and characterization of its cDNA from soybean and Vigna. *Plant J.* **7:**77–86.

187. Dunn, K., R. Dickstein, R. Feinbaum, B. K. Burnett, T. K. Peterman, G. Thoidis, H. M. Goodman, and F. M. Ausubel. 1988. Developmental regulation of nodule-specific genes in alfalfa root nodules. *Mol. Plant-Microbe Interact.* **1:**66–74.

188. Tingey, S. V., E. L. Walker, and G. M. Coruzzi. 1987. Glutamine synthetase genes of pea encode distinct polypeptides which are differentially expressed in leaves, roots and nodules. *EMBO J.* **6:**1–10.

189. Thummler, F. and D. P. S. Verma. 1987. Nodulin 100 of soybean is the subunit of sucrose synthase regulated by the availability of free heme in nodules. *J. Biol. Chem.* **262:**14730–14736.

190. Gordon, A. J. 1991. Enzyme distribution between the cortex and the infected region of soybean nodules. *J. Exp. Bot.* **42:**961–967.

191. Kouchi, H. and S. Hata. 1995. GmN56, a novel nodule-specific cDNA from soybean root nodules encodes a protein homologous to isopropylmalate synthase (LeuA) and homocitrate synthase (NifV). *Mol. Plant-Microbe Interact.* **8:**172–176.

192. Horvath, B., R. Heidstra, M. Lados, M. Moerman, H. P. Spaink, J. C. Prome, A. Van Kammen, and T. Bisseling. 1993. Lipo-oligosaccharides of NI induce infection-related early nodulin gene expression in pea root hairs. *Plant J.* **4:**727–733.

193. Minami, E., H. Kouchi, J. R. Cohn, T. Ogawa, and G. Stacey. 1996. Expression of the early nodulin, ENOD40, in soybean roots in response to various lipo-chitin signal molecules. *Plant J.* **9:** (in press).

194. Truchet, G., D. G. Barker, S. Camut, F. de Billy, J. Vasse, and T. Huguet. 1989. Alfalfa nodulation in the absence of *Rhizobium. Mol. Gen. Genetics* **219:**65–68.

195. de Billy, F., D. G. Barker, P. Gallusci, and G. Truchet. 1991. Leghaemoglobin gene transcription is triggered in a single cell layer in the indeterminate nitrogen-fixing root nodule of alfalfa. *Plant J.* **1**:27–35.

196. Bauer, P., T. Coba de la pena, F. Frugier, S. Poirier, H. I. Mckhann, P. Ratet, S. Brown, M. Crespi, and A. Kondorosi. 1995. Role of plant hormones and carbon/nitrogen metabolism in controlling nodule initiation on alfalfa roots. In *Nitrogen Fixation: Fundamentals and Applications,* ed. I. A. Tikhonovich et al., pp. 443–448. Kluwer Academic Publishers, Netherlands.

CHAPTER

3

Avirulence Genes

Jan E. Leach and Frank F. White

INTRODUCTION

Plants of today have survived the ebb and flow of countless disease epidemics. The survival of plants depends on their ability to resist marauding pathogens, while at the same time, a pathogen's success hinges on its ability to either avoid recognition or to overcome the resistance mechanisms of the plant. This adaption process between plants and pathogens undoubtedly was accelerated in crop species by farmers who selected, either by choice or fate, those individual plants that best withstood the ravages of disease. In spite of the selection imposed by co-evolution, populations or individuals of the host exist that do not resist the pathogen and populations of the pathogen exist that do not elicit resistance in the host. Genetic analyses of the variation within host and pathogen populations led to the formulation of the gene-for-gene hypothesis.[1,2] This hypothesis predicts that resistance (or incompatibility) is governed by the interaction of single, and for the most part, dominant host resistance (*R*) genes with corresponding pathogen genes called *avirulence* (*avr*) genes. Any *R* gene and *avr* gene combination that leads to resistance results in an incompatible interaction. Correspondingly, combinations that do not lead to resistance represent compatible host/pathogen interactions and provide the opportunity for disease.

The resistant responses of plants, which have been reviewed recently,[3,4] are diverse and include (1) phytoalexin biosynthesis; (2) cell wall reinforcement by deposition of callose, lignin, and hydroxyproline-rich glycoproteins; and (3) production of proteinase inhibitors; various enzymes such as lysozymes, chitinases, glucanases, and peroxidases; and other inhibitory proteins. Induction of some or all of these defensive responses is observed in most resistant interactions that usually, but not always, culminate in the hypersensitive response (HR). The HR is rapid cell death common to plants that have active resistance to pathogens.[5] Biochemical events that are associated with the HR are electrolyte loss from plant cells, activation of a K^+/H^+ exchange response (called the XR), calcium influx, increased lipoxygenase activity, and the production of active oxygen species (for review, see reference 6). Recent evidence suggests that the HR might be part of a programmed cell death response generated in plants after pathogen attack.[7–9] Most of the avirulence genes that have been characterized are involved in interactions that lead to elicitation of the HR.

AVIRULENCE GENES, RACE SPECIFICITY, AND HOST RANGE

Strains of a pathogen species may contain varying numbers of two general classes of avirulence genes (for review, see references 10,11). The first and most familiar type was identified by *intra*pathovar or *intra*species comparisons. Avirulence genes of this type were detected in members of a single pathogen species by inoculation to sets of *host differentials* (host plant lines or cultivars with different resistance genes). The combination of avirulence genes within a particular strain specifies the physiologic *race* of the pathogen.[12] Thus, in this case, the avirulence genes impose race specificity on a pathogen that is otherwise compatible on a given plant species. For example, *avrA*, the first avirulence gene cloned, confers on the soybean pathogen *Pseudomonas syringae* pv. *glycinea* the ability to elicit the HR only on cultivars of soybean with the *Rpg2* gene for resistance.[13]

The second type of avirulence genes was identified after *inter*pathovar or *inter*species transfer. This type of gene, sometimes referred to as a heterologous avirulence gene, was characterized first by Whalen and coworkers,[14] who hypothesized that, in addition to determining host range within a plant species (race specificity), avirulence genes may determine, in part, the host range of a pathogen to different plant species (pathovar specificity). The first of this type of avirulence gene, *avrRxv*, was cloned from the pepper pathogen *Xanthomonas campestris* pv. *vesicatoria*, which elicits the HR on bean.[14] When introduced into the bean pathogen *X.*

campestris pv. *phaseoli, avrRxv* conferred the ability to elicit resistance in bean in a race/cultivar specific manner. Interestingly, *avrRxv* not only conferred avirulence to *X. campestris* pv. *phaseoli* for particular bean cultivars, but also resulted in the induction of resistance by several *X. campestris* pathovars on their normally susceptible hosts (viz., pv. *glycines* on soybean, pv. *vignicola* on cowpea, pv. *holcicola* on corn, pv. *alfalfae* on alfalfa, and pv. *malvacearum* on cotton). Therefore, each of these hosts, then, may contain genes that confer recognition of *avrRxv*.[14,15]

Several additional avirulence genes, including one from a fungus (*PWL2*),[16] have been identified by interpathovar or interspecies transfer. Five avirulence genes from *P. syringae* pv. *tomato* (*avrA, avrD, avrE, avrPto,* and *avrRpt2*) have been shown to be recognized by soybean, which is not a host for the pathogen.[17–20] The *avrPpiA* gene from *P. syringae* pv. *pisi*[21] alters the virulence of *P. syringae* pv. *phaseolicola* to bean and *P. syringae* pv. *maculicola* to *Arabidopsis*[22] in a cultivar- or ecotype-specific manner. Two genes, *avrPphB* and *avrPphD* from *P. syringae* pv. *phaseolicola,* confer avirulence to *P. syringae* pv. *pisi* on pea.[23,24] The *pthA* from *X. citri* functions not only in pathogenicity to citrus but also as an avirulence gene in *X. campestris* pvs. *alfalfae* and *phaseoli*.[25]

The identification of avirulence genes by interspecific screening is consistent with the hypothesis that avirulence genes result in the activation of a resistance response in nonhost plants and may contribute to the limitations of the host range of a pathogen. However, with the exception of the fungal gene *PWL2*,[16] none of the genes identified have proved to be lone determinants of host range. Mutation of the endogenous *avrRxv* gene in *X. campestris* pv. *vesicatoria* did not allow that pathovar to become pathogenic to bean, although the mutant multiplied to higher final populations than the wild-type strain.[14] Similarly, inactivation of *pthA* from *X. citri* resulted in loss of the ability to elicit an HR on bean, but did not result in pathogenicity of *X. citri* to bean.[25] Inactivation of an avirulence gene from *Pseudomonas solanacearum* rendered the strain incapable of eliciting the HR on tobacco, and, although the mutant multiplied to levels only 50–fold less than those of a pathogenic strain, disease symptoms were not observed.[26] Multiple avirulence gene/resistance gene interactions may contribute to the resistance of the nonhost plants to these bacterial pathogens. However, a strain of *P. syringae* pv. *tomato* with mutations in each of four avirulence genes, which function to elicit the HR in soybean cultivars (*avrA, avrD, avrE,* and *avrPto*), was not pathogenic to soybean.[20] Although additional avirulence genes may condition the nonhost resistance, pathogens also may require positive gene functions that are necessary for pathogenicity on a nonhost plant species. Genes required for pathogenicity to different plant species, called host species specificity or *hsn* genes, have been described in several bacterial/plant systems.[25,27–29]

Some *h*ypersensitivity *r*eaction/*p*athogenicity (*hrp*) genes of gram-negative phytopathogenic bacteria, which are required for pathogenicity and the HR,[30–32] conceptually may be considered as avirulence genes. The *hrp* genes generally are clustered and encode elements of a transport system that is activated during in planta growth.[33] A group of proteins termed *harpins* (after the product of the *hrpN* gene of *Erwinia amylovora*[34]), which are thought to be transported out of the bacterial cell by the *hrp* mechanism, have been identified from three genera including *E. amylovora*,[34] *P. syringae*,[35] and *P. solanacearum*.[36] Harpins elicit a characteristic HR but are not referred to as avirulence genes because, in general, they have not been shown to confer cultivar specific responses. However, the harpin produced by the *popA1* gene of *P. solanacearum* elicits resistance only on certain lines of *Petunia hybrida*[36] and, therefore, could be considered a cultivar-specific elicitor similar, at least in function, to those directed by avirulence genes.

CHARACTERISTICS OF AVIRULENCE GENES

After characterization of more than 30 bacterial and fungal avirulence genes, common attributes of these genes and their products can be summed up as follows: 1. Sequence analysis reveals relatively little similarity among the various avirulence genes or to other sequences in the databases. 2. The genes generally encode single proteins and are induced in planta. 3. The products of the avirulence genes have not been detected outside the bacterial cell. Although these points are general observations, each has exceptions that are discussed next. Some viral proteins also function to elicit resistance in plants that contain single gene resistance and therefore, viral genes can be considered avirulence genes. Viral avirulence genes are the topic of chapter 6 and are not covered in detail here. Because of the greater complexity of fungal genomes, few fungal avirulence genes have been characterized.[11] Therefore, the bulk of the observations discussed herein come from the examination of avirulence genes in bacteria. Whether fungal genes are as enigmatic as bacterial genes remains to be seen.

Sequence Relationships among Avirulence Genes

Functional homologues to bacterial avirulence genes that have the same host specificity are found in different pathovars of a pathogen species. Functional alleles have been described for the *P. syringae* pv. *glycinea* genes *avrA* (in *P. syringae* pv. *tomato*[17]) and *avrC* (*avrPphC* from *P. syringae* pv. *phaseolicola*[37]), the *P. syringae* pv. *tomato* gene *avrD* (from *P. syringae* pvs. *lachrymans* and *phaseolicola*[38]), and the *P. syringae* pv. *pisi* gene *avrPpiA* (*avrRpm1* from *P. syringae* pv. *maculicola*[22]) (Table 3–1). Sequence similarity also exists between a few avirulence genes with different specificities. The deduced proteins of the *avrB* and *avrC* genes from *P. syringae* pv. *glycinea*, for example, share 42% amino acid identity.[39] Functional avirulence genes were created by the exchange of different domains from *avrB* and *avrC;* avirulence specificity was determined by the central region.[39] The only example of sequence similarity from avirulence genes in

Table 3-1. Avirulence Genes from Bacteria and Fungi

Pathogen	Avirulence Gene	Interacts with Plant (R gene)a	Alleles or Related Gene (+/- avirulence activity)	Regulation of Avirulence Genes	Size and Location of Avirulence Gene Productb	Other Known Functions	References
Pseudomonas syringae pv. *glycinea*	*avrA* (*avrPgyA1.Rpg2*)	Soybean (*Rpg2*)	Allele from *P. syringae* pv. *tomato* (+)	Requires *hrp* genes	101 kDa; no leader sequence or hydrophobic stretches	Aggressiveness to tomato	13, 148, 17
	avrB (*avrPgyB1.Rpg1*)	Soybean (*Rpg1*); *Arabidopsis* (*RPM1/RPS3*)	Sequence similarity to *avrC*	Repressed in rich media; induced in plants; requires *hrpSR* and *hrpL* genes; promoter has "*harp*-box"	36 kDa; no leader sequence or hydrophobic stretches		54,39,149, 55,56,150
	avrC (*avrPgyC1.Rpg3*)	Soybean, *Rpg3*	*avrPphC* from *P. syringae* pv. *phaseolicola* (+); similarity to *avrB*		39 kDa; no leader sequence or hydrophobic stretches		54,39,149,37
Pseudomonas syringae pv. *tomato*	*avrD* (*avrPtoA1*)	Soybean (*Rpg4*); soybean, bean	Alleles from *P. syringae* pvs. *glycinea* (-), *lachrymans* 1 (+), *lachrymans* 2 (+)	σ^{54} requiring promoter; regulated by *hrpS* & *hrpL*; repressed in rich medium, induced in plants; promoter has "*harp*-box"	34 kDa; no leader sequence; no hydrophobic stretches	Mutation does not affect aggressiveness to tomato	17,51,150,152, 38,135,18, 53,137
	avrE (*avrPtoE1*)	Soybean		Requires *hrp* genes		Aggressiveness to tomato	53,20
	avrPto (*avrPtoC1*)	Tomato, *Pto*		Highest expression in MM + sugars or alcohols; lowest in MM + TCA intermediates and rich media; induced in plants; requires *hrpSR;* promoter has "*harp* box"	18.3 kDa; highly hydrophilic; first 6 aa are hydrophobic		19,57

Table 3-1. *(con't.)* Avirulence Genes from Bacteria and Fungi

Pathogen	Avirulence Gene	Interacts with Plant (R gene)[a]	Alleles or Related Gene (+/- avirulence activity)	Regulation of Avirulence Genes	Size and Location of Avirulence Gene Product[b]	Other Known Functions	References
	avrRpt2 (*avrPtoB1*)	*Arabidopsis* (*RPS2*); soybean, turnip, radish		Highly expressed in minimal medium; low in rich medium; dependent on *hrpRS* locus; promoter has "*harp* box"	28.2 kDa; hydrophilic, no signal peptides		153,55,154, 155
Pseudomonas syringae pv. *pisi*	*avrPspil* (*avrPpiA1.R2*)	Pea (*R2*); *Arabidopsis, RPM1;* bean, *RN* + *RN2;* soybean	*avrRPM1* (*avrPmaA1.RPM1*) from *P. syringae* pv. *maculicola* 1 (+); *maculicola* 2 (+)			*avrRpm1* confers aggressiveness to *Arabidopsis*	21,23,22, 156,10
	avrPpi3 (*avrPpiB.R3*)	Pea, *R3*					157,158
Pseudomonas syringae pv. *phaseolicola*	*avrPphl* (*avrPphA1.R1*)	Bean, *R1*					159
	avrPphB (*avrPph3.R3*)	Bean (*R3*); pea	No hybridizing sequences in virulent races				23
	avrPphE (*avrPph2.R2*)	Bean (*R2*); pea	Hybridizes to virulent races		41 kDa; hydrophilic; no membrane spanning domains		71,59
	avrPphD	Pea					24

Table 3-1. *(con't.)* Avirulence Genes from Bacteria and Fungi

Pathogen	Avirulence Gene	Interacts with Plant (R gene)[a]	Alleles or Related Gene (+/- avirulence activity)	Regulation of Avirulence Genes	Size and Location of Avirulence Gene Product[b]	Other Known Functions	References
Pseudomonas solanacearum	*avrA*	Tobacco				Host species resistance	26
Xanthomonas campestris pv. *vesicatoria*	*avrRxv*	Bean (*Rxv*); soybean, cowpea, alfalfa, cotton, corn					14,15
	avrBs1	Pepper, *Bs1*	Similarity to *avrA* from *P. syringae* pv. *glycinea*		49.8 kDa; no hydrophobic stretches or leader sequences		160,40,108
	avrBs2	Pepper, *Bs2*	Functional homologues in *X. campestris* pvs. *alfalfae, malvacearum, vignicola, vitians, campestris,* and *phaseoli; X. oryzae* pv. *oryzae?*		60 kDa	Aggressiveness to pepper and alfalfa; not to rice	100,99,101
	avrBs3	Pepper, *Bs3*	Related to genes from *X. campestris* pv. *vesicatoria, X. oryzae* pv. *oryzae, X. campestris* pv. *malvacearum, X.citri*	Expression does not require *hrp* gene expression; constitutive	122 kDa; 17.5 102–bp repeats; cytoplasmic		42
	avrBsT	Pepper, *BsTi*					100
	avrBsP (*avrBs3–2*)	Tomato	*avrBs3* homologue	Constitutive	122 kDa; 17.5 102–bp repeats		44,47

Table 3-1. (*con't.*) Avirulence Genes from Bacteria and Fungi

Pathogen	Avirulence Gene	Interacts with Plant (R gene)[a]	Alleles or Related Gene (+/- avirulence activity)	Regulation of Avirulence Genes	Size and Location of Avirulence Gene Product[b]	Other Known Functions	References
Xanthomonas campestris pv. *malvacearum*	*avrB4*	Cotton, *AcB1, AcB4*	*avrBs3* homologue		19 102–bp repeats		45,48
	avrb6	Cotton, *AcB1, Acb6*	*avrBs3* homologue		13.5 102–bp repeats	Aggressiveness to cotton	45,48
	avrb7	Cotton, *Acb7*	*avrBs3* homologue		19 102–bp		45,48
	avrB101	Cotton	*avrBs3* homologue		23 102–bp repeats		45,48
	avrBln	Cotton, *AcBln*	*avrBs3* homologue				45,48
	avrB102	Cotton					45,48
Xanthomonas campestris pv. *citri*	*pthA*	Bean, alfalfa	*avrBs3* homologue			Pathogenicity to citrus	102,25
Xanthomonas campestris pv. *raphani*	*avrXca*	*Arabidopsis* (most ecotypes); not *Brassica*	None			No visible HR	161
Xanthomonas oryzae	*avrXa7*	Rice, *Xa-7*	*avrBs3* homologue	Constitutive	125 kDa, cytoplasmic; 25.5 102–bp repeats	Aggressiveness to rice	46,103
pv. *oryzae*	*avrXa10*	Rice, *Xa10*	*avrBs3* homologue		119 kDa, cytoplasmic; 15.5 102–bp repeats		46

Table 3-1. (*con't.*) Avirulence Genes from Bacteria and Fungi

Pathogen	Avirulence Gene	Interacts with Plant (R gene)[a]	Alleles or Related Gene (+/- avirulence activity)	Regulation of Avirulence Genes	Size and Location of Avirulence Gene Product[b]	Other Known Functions	References
	avrxa5	Rice, *xa-5*	*avrBs3* homologue				46
Cladosporium fulvum	*Avr4*	Tomato, *Cf4*	Alleles from virulent races are not elicitor active		Pre-pro-protein, 135 aa; pro-protein, 124 aa; mature elicitor, 106 aa, is extracellular		64
	Avr9	Tomato, *Cf9*	No homologous DNA in virulent races	Induced in plants; induced in low nitrogen medium (promoter similar to *nit-3* gene)	Pre-pro-protein, 63 aa; pro protein, 40 aa; mature elicitor, 28 aa, is extracellular		75,106,131, 104
Magneporthe grisea	*PWL2*	Weeping lovegrass	*PWL1* (+), *PWL3* (-), & *PWL4* (+)		145 aa; glycine- rich; signal peptide		16
	AVR2–YAMO	Rice			223 aa	Possible neutral Zn-protease	16
Rhynchosporium secalis	Putative *avrRrs1*	Barley, *Rrs1*	Inactive alleles from virulent race		Precursor = 82 aa/ mature NIP1 = 60 aa (10 cysteine residues)	Aggressiveness to barley	78,77
Phytophthora parasitica	Putative *parA1*	Tobacco; radish, turnip	Alleles in producers and nonproducers		Preprotein of 118 aa; mature parasticein elicitor = 98 aa (6 cys)	Aggressiveness to tobacco	82,83,84

[a]The first plant (R gene, where known) is the plant on which the avirulence gene was identified. Others also exhibit a resistant response in interactions with the avirulence gene.

[b]Based on DNA sequence data, in most cases.

different bacterial genera is provided by *avrBs1* from *X. campestris* pv. *vesicatoria* and *avrA* from *P. syringae* pv. *glycinea,* which share 47% sequence identity in their carboxy terminal regions.[40]

In a few cases, avirulence genes share sequence similarity with other genes in the data bases. The *X. campestris* pv. *vesicatoria* gene *avrRxv* shows similarity to the *yopJ* gene of *Yersinia pseudotuberculosis*.[41] This relationship is particularly interesting, because the *yopJ* product is secreted and components of the *Yersinia* YOP secretion pathway are related to the *hrp* pathway.[33] The location of the product of *avrRxv* has not been reported. The deduced amino acid sequence of the fungal avirulence gene *AVR2–YAMO* from *Magneporthe grisea* contains short stretches that share identity with the active center of a neutral Zn-protease.[16] The significance, if any, of these similarities is not known.

The situation for *avrBs3* is entirely different from other bacterial avirulence genes with regard to sequence identity. The *avrBs3* gene is the first-identified member of a large gene family and was cloned from the pepper and tomato pathogen, *X. campestris* pv. *vesicatoria.*[42] This gene family, which encodes proteins with 90–97% amino acid sequence identity[43] among family members, is distributed widely in several *Xanthomonas* pathovars.[25,42,44–47] The genes are found exclusively in xanthomonads and have been identified in pathogens of diverse dicots, such as tomato,[44,47] cotton,[45] and citrus,[25] and monocots such as rice.[46] Note that not all avirulence genes from xanthomonads are members of the *avrBs3* gene family; *avrRxv, avrBs1,* and *avrBs2* from *X. campestris* pv. *vesicatoria* share no apparent sequence similarity with *avrBs3* (Table 3–1).

Within the coding region of each *avrBs3*-homologue is a series of 102–bp, directly repeated, DNA sequences[25,42–47] (Fig. 3–1). Different *avrBs3*-homologues may contain different numbers of the 102–bp repeat (Table 3–1). The *avrBs3, avrBs3–2,* and *pthA* genes all contain 17.5 copies,[25,42,47] *avrXa10* contains 15.5 copies,[46] and *avrB6* contains 13.5 copies.[45] When repeat domains were exchanged between avirulence genes, the ability to elicit resistance on a particular host differential was dependent on the presence of the corresponding repeat domain in the avirulence gene (e.g., *avrXa7* repeat domain was required for avirulence on rice with *Xa-7*).[48,49] Thus, the specificity of the *avrBs3*-like genes for particular resistance genes resides within the repeat domain.

Within the same *avrBs3*-gene homologue, the 102–bp repeats share 91–100% identity with each other.[25,42,44–47] Most of the differences between repeats are located within a two-codon region referred to as the *variable region.*[46] The codons and predicted amino acids in the variable region vary. Although *avrBs3, avrBsP,* and *avrXa10* have some of the same two-codon combinations in the variable regions of their 102–bp repeats, the order in which they are arranged in each gene is different (Fig. 3–1B). Herbers et al.[50] and Bonas et al.[47] demonstrated that the number of repeats present in the *avrBs3* homologue is not sufficient by itself to dictate specificity and the organization or number of specific repeats (i.e., with

A.

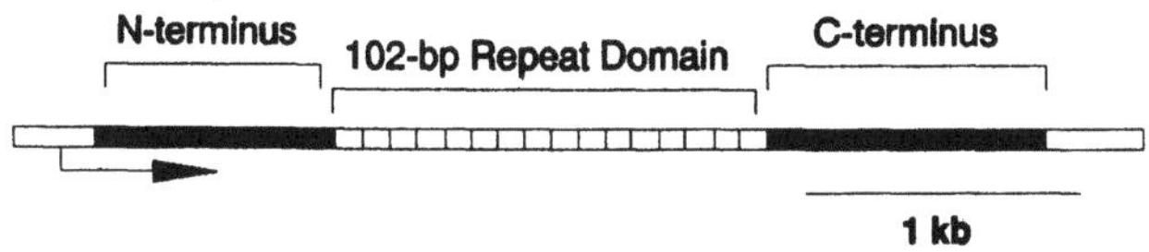

B.

	1	2	3	4	5	6	7	8	9	10	11	12	13	14	15	16	17	18
avrBs3	HD	NG	NS	NG	NI	NI	NI	HD	HD	NG	NS	NS	HD	HD	HD	NG	HD	NG
avrBs3-2	NI	NG	NI	NI	NG	NG	NI	NS	NG	NI	NS	NG	HD	HD	NS	HD	NG	NG
avrXa10	NI	HG	NI	HG	NI	NI	NN	HD	NI	HD	NN	HG	NS	NG	HD	NG		
avrb6	HD	NI	NG	HD	HD	NI	HD	NI	NS	HD	HD	HD	NN	NG				
Repeat #	1	2	3	4	5	6	7	8	9	10	11	12	13	14	15	16	17	18

Figure 3-1. *A.* Schematic map of members of the *avrBs3* gene family. The filled bars indicate the coding sequence on either side of the repeat domain, which consists of varying numbers of 102-bp directly repeated sequences. *B.* The order of the two-amino acid variable region from within each 34 amino acid repeat of *avrBs3*[42] and *avrBs3-2*[47] from *Xanthomonas campestris* pv. *vesicatoria, avrXa10*[46] from *X. oryzae* pv. *oryzae*, and *avrb6*[45] from *X. campestris* pv. *malvacearum.*

particular variable regions) is likely responsible for cultivar specificity. Random deletion of 102–bp repeats from *avrBs3* created mutations with different numbers of repeats.[50] Some, but not all, deletions abolished *avrBs3* activity when tested on pepper with *Bs3,* indicating that the number of repeats was not critical for activity. However, other deletions of the same number of different specific repeats retained avirulence activity. A very interesting finding was that some deletions created new specificities.[50] Some clones with deletions of *avrBs3* repeats lost avirulence activity on pepper with *Bs3* but gained avirulence activity on the pepper cultivar without *Bs3*. Other deletions gained avirulence activity toward an *avrBs3*-compatible tomato cultivar. Thus, the new specificities (avirulence genes) created by these deletions uncovered unknown resistance genes in both hosts.

Another puzzle in the structural analysis of this fascinating gene family is that deletion of the carboxy terminus of the protein from *avrBsP,* a homologue of *avrBs3* from a *X. campestris* pv. *vesicatoria* strain pathogenic on pepper and avirulent on tomato,[44] does not affect avirulence activity, whereas deletion of that portion of the protein from *avrBs3* results in a loss of activity.[42,47] The *avrBsP* gene has the same amino terminal region as *avrBs3* but, as originally cloned, only six copies of the repeat sequence and none of the carboxy terminal region. The gene apparently was isolated as a truncated version of *avrBs3–2*, which subsequently was cloned and shown to have 17.5 copies of the repeat. Like *avrBsP,* the *avrBs3–2* gene confers the ability to induce the HR in tomato but not pepper[47] and can be reduced to only three repeats while retaining avirulence activity.[42,47]

Avirulence Gene Structure and Organization

The majority of avirulence genes appear to be monocistronic. However, genes *avrC*[39] and *avrD*[51] are encoded by the first open reading frame (ORF) of two and five ORF operons, respectively. None of the other ORFs are required for avirulence activity. More recently, two possible exceptions to the single ORF rule have been reported. The activities of *avrE* from *P. syringae* pv. *tomato*[52,53] and *avrPphD* from *P. syringae* pv. *phaseolicola*[24] appear to require two ORFs. Many avirulence genes are bounded by repetitive elements.[25,45,47,54,55] Several are located on plasmids (for review, see Tables 1 and 2 in reference 10), such as the six avirulence genes of the *avrBs3* family in *X. campestris* pv. *malvacearum*, which are located on a plasmid.[45] The *avrB* gene has been postulated to have been derived recently from outside the *P. syringae* group because the codon usage does not reflect the GC content of the genome.[39]

Regulation of Avirulence Genes

The expression of many avirulence genes from *P. syringae* is inducible in planta or by growth in minimal media with optimal expression at pH 5.5.[55–61] For example, expression of the *avrRpt2* gene from *P. syringae* pv. *tomato* was highest in minimal medium containing citrate or citrate and fructose as carbon sources and was not detected in cells grown in King's medium B, a complex medium.[55] In general, the conditions for best avirulence gene expression are similar to the environment encountered by *P. syringae* in the leaf apoplast, which is typically about pH 5.5[62] and limited in nutrients.[63] Both *Avr4* and *Avr9* from the fungal pathogen *Cladosporium fulvum* also are induced in planta, and *Avr9* is induced under nitrogen-limiting conditions.[64,65] The regulation of many avirulence genes is consistent with a role in plant-pathogen interactions other than the elicitation of resistance. The exceptions to these observations are again members of the *avrBs3* gene family; these genes appear to be expressed constitutively in the pathogen.[47,66,67]

Although *hrp* gene function is required to manifest the avirulence phenotype in plants, mutations in most *hrp* loci do not affect *P. syringae* avirulence gene expression.[55–57] However, at least the *hrpSR* and *hrpL* loci are required for expression of *avrB, avrD, avrE, avrRpt2,* and *avrPto*.[52,55,57,58] In *P. syringae* pv. *glycinea, hrpR* activates transcription of *hrpS*.[68] The *hrpS* gene has sequence similarity to the σ^{54}-requiring *ntrC* family of transcriptional activators and includes a putative DNA binding domain.[69] Interestingly, the *avrD* promoter contains a σ^{54} binding site, and the σ^{54} gene, *ntrA,* is required for expression of *avrD*.[58] No other *P. syringae* avirulence gene promoters contain the consensus sequence of the σ^{54}-binding site, although *P. syringae* avirulence genes and several *hrp* genes contain a common sequence motif called the harp box (consensus 5'T/G-G-G-A-A-C-C-$N^{15,16}$-C-C-A-C).[22,55,57–59,70,71] This presumptive promoter element is not similar to consensus promoters of *Escherichia coli* genes that are transcribed by either

σ^{54}- or σ^{70}- associated RNA polymerases or to other *Pseudomonas* promoters.[55] Recently, Xiao et al.[72] suggested that, in *P. syringae* pv. syringae, the HrpR and HrpS proteins may function as a dimer to activate the transcription of *hrpL*. HrpL alone is sufficient to activate other *hrp* and possibly avirulence gene promoters. HrpL replacement for σ^{54} could explain the lack of typical σ^{54} promoters upstream of some of the *P. syringae* avirulence genes. Transcription of the *avrBs3*-type genes of *X. campestris* pv. *vesicatoria* is not dependent on *hrp* gene activity, although they do require *hrp* gene function for the avirulence gene phenotype. The specific expression of some avirulence genes during in planta growth and in concert with the *hrp* regulon is consistent with a role for the genes in some aspect of the host/pathogen interaction other than the elicitation of resistance.

Location of Avirulence Gene Products

Only fungal avirulence gene products have been detected extracellularly.[11] Modified products of the *Avr4* and *Avr9* genes accumulate in intercellular fluids of tomato leaves during compatible interactions between races of *Cladosporium fulvum* with *Avr4* and *Avr9* activities and a susceptible host (without the *Cf4* or *Cf9* genes for resistance).[73–76] Other fungal avirulence gene products that are located extracellularly are *n*ecrosis-*i*nducing *p*eptide (NIP1), thought to be the product of the *Rhynchosporium secalis* avirulence gene *avrRrs1*,[77,78] and the elicitins from *Phytophthora* spp.[79–84]

With one exception (the *avrD*-directed syringolide elicitors),[18] no race-specific elicitor has been isolated from a bacterial pathogen or from plant tissues during incompatible interactions with bacteria. The inability to detect extracellular race-specific elicitors in bacterial interactions is puzzling, particularly for the *avrBs3* gene family, where the structural requirements for avirulence activity suggest that the products of these genes are direct elicitors. Yet, the evidence to date indicates that the protein products of *avrBs3*-homologues are not extracellular elicitors. The predicted proteins are hydrophilic and do not contain a signal peptide sequence.[25,42,44–47] The products from both *avrBs3* (from *X. campestris* pv. *vesicatoria*) and *avrXa10* (from *X. oryzae* pv. *oryzae*) have been shown by biochemical fractionation and immunoelectron microscopy experiments to be located mainly in the cytoplasm of the bacterial cells.[66,67,85] In both cases, the avirulence genes were overexpressed to increase the amount of detectable antigen. Young et al.[67] used not only epitope-tagged *avrXa10* for specific detection of the gene product with monoclonal antibodies, but also a polyclonal antibody that was shown to bind to the repeat domain of the protein. Thus, even if only portions of the repeat domain were exported, the peptides should have been detected in either concentrated culture filtrates or by immunolabeling of infected plant tissues. Finally, neither the *avrXa10* gene product, which was purified from *E. coli* or *X. oryzae* pv. *oryzae* cells, nor the extracellular fluids of rice leaves undergoing an incompatible response elicited the HR in rice plants containing *Xa-10*.[67]

The evidence that these avirulence gene products do not function outside the bacterial cell, although compelling, remains inconclusive. Bacterial proteins can be exported through alternate secretion pathways without the need for signal sequences,[33] and some exported proteins, such as α-hemolysins, require modification for activity.[86,87] Thus, the avirulence gene products may not have been detected outside the bacterial cells because they are: (1) modified such that they are not recognized by the antibodies; (2) rapidly degraded by host proteases, as found for some harpins;[34] (3) exported in very low amounts and/or at very specific times during infection, or (4) introduced by some mechanism directly into the plant cells. Precedence for direct introduction to eukaryotic cells can be found in secretory systems of animal pathogens. *Yersinia pseudotuberculosis* exports the cytotoxic product of the *yopE* gene directly to HeLa cells during infection of the cells with bacteria.[88] Curiously, the export mechanism of the *yopE* cytotoxin is related to the *hrp* secretory apparatus.[33,89–92]

An intriguing recent report of Yang and Gabriel[93] provides evidence that avirulence genes contain sequences that function inside plant cells. They identified sequences resembling three nuclear localization signals (NLSs) in the carboxy-terminal regions of the predicted proteins of all sequenced members of the *avrBs3*-family. When introduced into onion cells, β-galactosidase translational fusions with DNA encoding the C-terminal regions of two *avrBs3* homologues (*avrb6* and *pthA*) were transiently expressed, and GUS activity was detected specifically in the nuclei of transformed cells. Yang and Gabriel[93] also found heptad repeats similar to leucine zippers in each of the *avrBs3*-homologue sequences. Leucine zippers and leucine-rich repeats may serve as sites for protein-protein interactions.[94] One possible model from this data is that the avirulence gene product is delivered to the plant cell, where it could interact directly with resistance gene products. The presence of NLSs in a bacterial gene does not prove they function in nature to localize the bacterial protein to the plant nucleus. In fact, deletion of the region encoding most of the C-terminus (including the NLSs) from one *avrBs3*-homologue, *avrBs3–2*, does not affect avirulence activity.[42,47] Although in no way conclusive, these intriguing findings may require that we revisit our thinking with regard to avirulence gene product location and function.

DO AVIRULENCE GENES HAVE OTHER FUNCTIONS?

Because inactivation of the avirulence genes in the pathogen in several cases does not attenuate the pathogen's ability to colonize and cause symptoms on the host plant (for review, see reference 95), and no enzymatic or physiological function can be ascribed to most avirulence gene products, avirulence genes generally were not thought to be critical for the survival of the pathogen.[95,96] The lack of any contribution to the fitness of the pathogen produced the conundrum: Why would a gene, whose only apparent function is deleterious, be maintained in an organism? The logical conclusion is that avirulence genes must serve some additional functions in the pathogen, and the conditions under which the phenotypes can be observed have not been identified.

At least one protein with avirulence activity has been identified for which a specific biochemical function apart from avirulence is known. Single amino acid substitutions in the coat protein of Tobacco Mosaic Virus resulted in virus that caused the induction of a localized HR in tobacco carrying the *N'* gene.[97,98] All mutants caused systemic, mosaic symptoms (wild-type) on the susceptible (*n'*) genotype tobacco. Therefore, the coat protein serves both as the structural protein for encapsulating the viral RNA and an elicitor of resistance.

Further evidence, although less direct, supports the hypothesis that avirulence gene products have other functions. Kearney and Staskawicz[99] noticed that the pepper *Bs2* gene, which confers resistance to the pathogen *X. campestris* pv. *vesicatoria,* had not been overcome in the field. Sequences related to *avrBs2,* the avirulence gene corresponding to *Bs2,* occur in all *X. campestris* pv. *vesicatoria* strains and in many other *X. campestris* and *X. oryzae* pathovars.[99–101] Strains of *X. campestris* pv. *vesicatoria* containing an inactivated *avrBs2* gene were less aggressive to a susceptible cultivar of pepper than the wild-type strain.[99] Similarly, *avrBs2* mutants of *X. campestris* pv. *alfalfae* were reduced in their ability to multiply in alfalfa plants. Complementation of the mutant with a plasmid-borne copy of *avrBs2* restored growth to wild-type levels. Therefore, *avrBs2* appears to be essential for the growth of the pathogen in these hosts. The durability of *Bs2* for control of bacterial spot in pepper, in fact, may be a consequence of the requirement for *avrBs2* for overall fitness of the pathogen. Interestingly, mutants in the *avrBs2* homologue in the rice pathogen, *X. oryzae* pv. *oryzae,* did not alter the aggressiveness of this pathogen to rice.[101] *X. oryzae* pv. *oryzae* invades the vascular system of rice, whereas *X. campestris* pvs. *vesicatoria* and *alfalfae* colonize the spaces between their host's mesophyll cells. In interactions involving vascular pathogens such as *X. oryzae* pv. *oryzae,* the *avrBs2* homologue may not be required for colonization of vessels, or alternatively, the phenotype the gene confers may be more subtle.

Other avirulence genes that function in aggressiveness or fitness to host plants are the *avrE* and *avrA* genes from *P. syringae* pv. *tomato,*[20,53] *avrRpm1* from *P. syringae* pv. *maculicola,*[10] *pthA* gene from *X. citri,*[102] *avrb6* from *X. campestris* pv. *malvacearum,*[45] and *avrXa7* from *X. oryzae* pv. *oryzae.*[103] The *pthA* gene, which functions as an avirulence gene on bean and cotton (when present in *X. campestris* pvs. *phaseoli* and *malvacearum,* respectively), originally was identified as a gene for pathogenicity in *X. citri* and is required for induction of Asiatic citrus canker symptoms and colonization of citrus.[102]

The fact remains that inactivation of many bacterial avirulence genes, such as *avrD*,[20] has produced no measurable effects on the ability of the bacteria to multiply, colonize, or evoke symptoms in plants under experimental conditions. Mutations in the fungal avirulence gene *Avr9* similarly did not impair pathogenicity of *C. fulvum* to tomato, suggesting that the gene is dispensable to the pathogen.[104] De Wit[11] noted that the corresponding resistance gene *Cf9* has provided complete protection to commercial tomato lines since its introduction in 1979. The stability of *Cf9,* like that of *Bs2,* suggests that *Avr9* is essential to the pathogen under field conditions. Perhaps avirulence genes like *Avr9* contribute a function as yet unidentified, such as survival during environment stress (e. g., protection against desiccation, solar irradiation, and osmotic or pH extremes); effectiveness at the infection court; or promoting the secondary spread of the pathogen.

HOW DO PATHOGENS AVOID AVIRULENCE IN NATURE?

In plant-pathogen systems, evolution of a pathogen population is directed toward avoiding recognition and induction of resistance by the host plants. Avoidance may result from the masking of elicitation of resistance by production of aggressiveness factors. Pectic enzyme production by *Erwinia chrysanthemi* masks expression of the HR.[105] Alternatively, lack of recognition may be due to modification or absence of the avirulence gene. The identification and characterization of avirulence genes and their alleles have allowed for clarification of the genetic mechanisms that a pathogen uses to avoid recognition. In some cases, alleles to the avirulence genes are not present in virulent races. Races of *C. fulvum* virulent to tomato with *Cf9,* for example, do not contain DNA that hybridizes with *Avr9* gene and, therefore, are not recognized by the *Cf9* resistance gene.[106] Alternatively, DNA that hybridizes to the avirulence genes may be present, but the alleles do not confer avirulence activity. Lack of activity may be due to minor changes in the sequences of avirulence genes that alter function. Sequence analysis of inactive alleles of avirulence genes such as *Avr4* from *C. fulvum*[64] revealed that single base-pair changes were responsible for the lack of avirulence function on tomato with *Cf4.* Comparison of sequences from five alleles of *avrD* from pathovars of *P. syringae* other than *P. syringae* pv. *tomato* permitted the identification of four amino acid substitutions that were unique to a nonfunctional allele from *P. syringae* pv. *glycinea.*[38] More radical changes can involve insertions by mobile DNA elements. Spontaneous mutations to virulence on pepper with *Bs1* were detected at frequencies of 5×10^{-6} in strains of the bacterial spot pathogen of pepper (*X. campestris* pv. *vesicatoria*) carrying *avrBs1.*[107] Examination of several mutants revealed that an insertion element had transposed to the *avrBs1* locus and inactivated the gene.[108]

In cases where a multigene family of avirulence genes exists within a strain, intragenic or intergenic recombination could account for divergence between the copies.[109] New host specificities of the *avrBs3* homologue from *X. campestris* pv. *vesicatoria* were created in vitro by changing the number of 102–bp repetitive sequences.[47,50] Presumably, similar reassortment can occur by recombination in vivo. In *X. campestris* pv. *malvacearum,* the cotton pathogen, spontaneous race-change mutations occur at very high frequencies (10^{-3}–10^{-4}).[110] Most of the avirulence genes characterized from this pathogen are members of the *avrBs3*-gene family, and, as such, are related by a high degree of sequence identity. Characterization of five race-change mutants by Southern analysis revealed size changes at the appropriate avirulence locus, suggesting that either recombination between genes or insertional inactivation might have occurred.[45] Comparison of sequences between these mutant genes and their flanking regions would be necessary to reveal recombinational breakpoints to prove intragenic recombination.

PHENOTYPE OF THE RESISTANCE RESPONSES

The HR determined by different avirulence gene/resistance gene interactions within the same pathogen/host interaction can vary phenotypically. The HR governed by interactions between *avrPphE* in *P. syringae* pv. *phaseolicola* and bean with *R2* is macroscopically different from that observed in interactions between *avrPphB* and *R3*.[59] The more rapid HR determined by *R3* is associated with higher levels of phytoalexin accumulation at inoculation sites. In soybean, *avrB* in *P. syringae* pv. *glycinea* causes a more rapid response than *avrC*.[39,54] In interactions between *X. oryzae* pv. *oryzae* and rice, the timing and dynamics of events associated with resistance (increased activity of a cationic peroxidase, lignin deposition, host cell death, and a decrease in the rate of bacterial multiplication) are dependent on the specific avirulence gene/resistance gene interaction.[46,111] Responses associated with the HR in interactions involving the avirulence gene *avrXa7* and its corresponding resistance gene *Xa-7* occur later than in those in interactions between *avrXa10* and *Xa-10,* and the phenotypes of the HRs are distinct. Differences could be due to (1) differences in levels of expression of the avirulence genes, (2) different levels of activation of a common signal transduction pathway leading to the defense response, or (3) activation of different transduction pathways that lead to resistance.[59] For example, overexpression of the related avirulence genes (members of the *avrBs3*-gene family) from other species of *Xanthomonas* leads to more rapid or stronger HRs in host cultivars with the corresponding resistance genes.[48,85] Characterization of the molecules involved (elicitors and receptors) and analysis of their interactions will clarify how different avirulence gene/resistance gene interactions result in phenotypically different HRs.

MODELS FOR AVIRULENCE GENE FUNCTION

Several models for avirulence gene function have been described over the years to explain the molecular basis for induction of resistance in race-specific interactions.[112–116] A myriad of other models or variations of the models presented herein undoubtedly can be conceived. Of these, the elicitor/receptor models have received the most attention.[113] The simplest model predicts that the primary avirulence gene product is the elicitor of resistance (Model 1, Fig. 3–2). Alternatively, the elicitor may be a metabolite whose synthesis is catalyzed or otherwise regulated by the avirulence gene product (Model 2, Fig.3–2). In either model, the elicitor may bind extracellularly or intracellularly to a unique receptor, which may be the product of the resistance gene (as shown in Fig. 3–2), or a receptor common to both resistant and susceptible plants (not shown). In the second scenario, the resistance gene product may be a unique link in the signal transduction process, and the absence of the resistance gene may prevent transduction of the signal to the defense response pathway. A variation of the elicitor/receptor model is that the elicitor (either a protein or a metabolite) may be a toxinlike compound and binding by the receptor may direct the toxin to a site for insertion into the cell membrane or site of inhibition (or activation) of a host enzyme (not shown).

A third and more complex model proposes that the plant resistance genes direct the synthesis of a unique inducer compound (Model 3, Fig. 3–2). The inducer interacts with the corresponding avirulence gene product in the pathogen; this in-

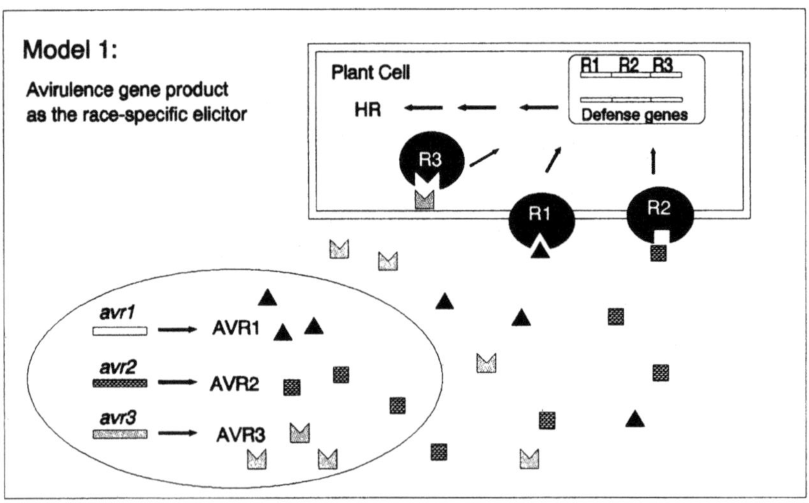

Figure 3-2. Models for avirulence gene function in the elicitation of plant defense responses.

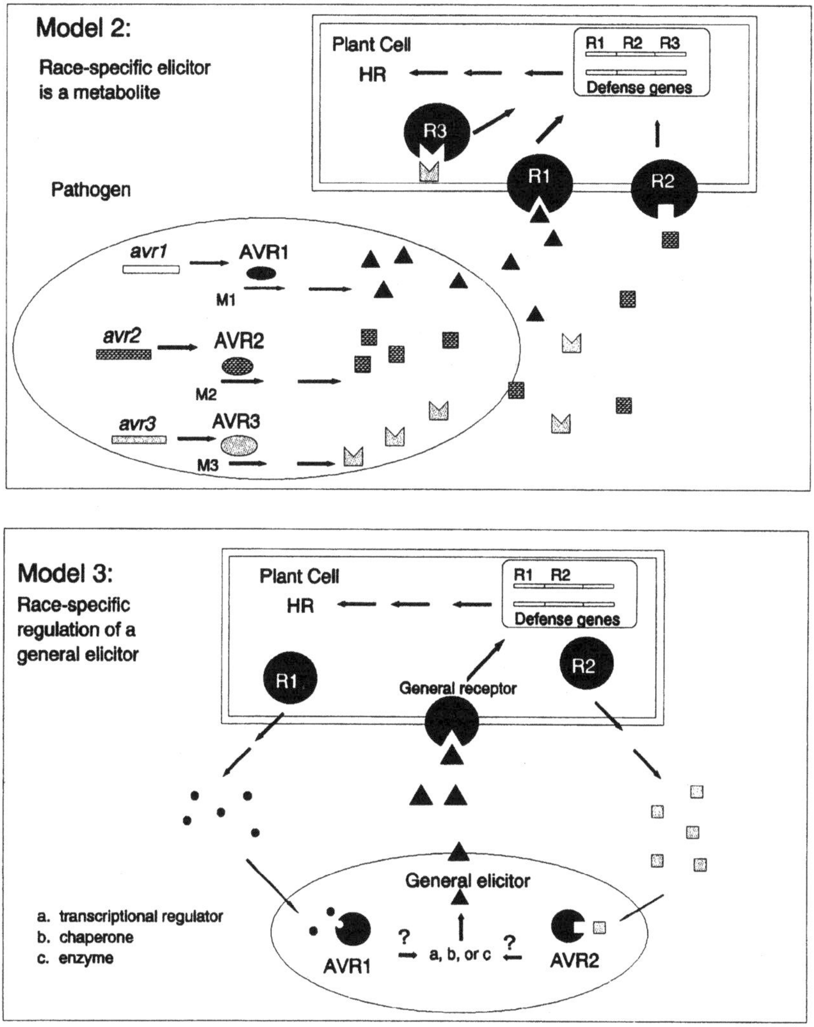

Figure 3-2. (*Con't.*)

teraction, in turn, regulates *production* or *export* of a *general elicitor* of resistance (Model 3, Fig. 3–1). In Model 3, the plant compound (derived by the resistance gene product) or a bacterial compound (perhaps a *hrp* gene product) may be metabolized by the avirulence gene product to a general elicitor. This model is based partially on *Rhizobium*[117] and *Agrobacterium*[118] interactions in which expression of bacterial genes in nodulation or virulence, respectively, are dependent on the perception of plant signals, usually flavonoids. For example, in the presence of appropriate plant flavonoids, regulatory NodD proteins activate the transcription of the *nod* and *nol* genes (for reviews see references 117,119–121). Several of the proteins encoded by the *nod* genes are involved in the biosynthesis of signal molecules that, in turn, mediate host specificity of nodulation.[121,122] Model 3 differs from other models in that at least one level of specificity results from the interaction of a plant signal compound with the avirulence gene product *inside the bacterium.*

Confirmation of any of the models awaits corroborative evidence from the molecular analysis of pathogen and plant components involved in elicitation of resistance. However, some information is available from the structure of a number of recently cloned resistance genes. The first resistance gene involved in a gene-for-gene type interaction to be cloned was *Pto,* which confers resistance to strains of the bacterium *P. syringae* pv. *tomato* that carry *avrPto.*[123] Cloning and characterization of four additional resistance genes have been reported recently [124–129] (for review, see reference 130). Significantly, common features are evident among the six cloned resistance genes (Table 3–2). If the six genes are indeed representative of resistance genes in general, the sequence data provide evidence that resistance is signaled in a manner similar to the better-characterized pathways observed in animal cells. Some of the plant resistance genes may have ancestral genes common to animal genes. The *N* [128] and *RPS2*[125,126] genes have similarity to a number of signal transduction proteins including Toll from *Drosophila,* which is involved in embryo development, and IL-1R, the interleukin receptor from humans, which, curiously is involved in disease defenses. Both have kinaselike domains and leucine-rich repeats (LRRs), which are thought to be involved in protein-protein interactions (for review see reference 131), in one polypeptide. On the other hand, *Pto*[123] consists solely of a kinase domain, and *Cf9*[124] consists solely of the LRR portion. *N*[128] and *Pto*[123] are likely to be cytoplasmic, whereas *Xa21*[129] and*Cf9*[124] are possibly extracellular. These molecules, either directly or as part of a protein complex, probably bind to an elicitor molecule (either Model 1 or 2, Fig. 3–1) or detect the binding of the elicitor to a common receptor. Albeit few, these cases are consistent with the idea that avirulence genes direct the synthesis of the elicitor molecules that interact with the resistance gene product in a signal transduction complex.

Avirulence Genes and Their Role in Resistance Induction

Our understanding of the function of avirulence genes in resistance has remained frustratingly incomplete. In contrast to the plant resistance genes for which common domains are observed among the first few genes from a variety of plants and involving interactions with three very different pathogens (bacteria, fungi, and virus), the information so far gleaned from the molecular

Table 3–2. Resistance Genes and Their Characteristics

Host Plant	Resistance Gene	Interacting *avr* Gene	Features	Possible Location of Protein	Similarity to Other Genes	Reference
Tomato	*Pto*	*avrPto*	Protein kinase	Cytoplasm/membrane bound	Kinases	123
	Cf9	*Avr9*	28–LRRs[a]	Trans-membrane		124
Tobacco	*N*	TMV replicase?	14–LRRs, Nucleotide binding site, Leucine zipper	Cytoplasm	LRRs	128
Arabidopsis	*RPS2*	*avrRpt2*	14–LRRs, Nucleotide binding site, Leucine zipper	Cytoplasm/trans-membrane	LRRs	125, 126
Flax	*L*[6]		2–LRRs, Nucleotide binding site	Cytoplasm	LRRs	127
Rice	*Xa21*		23–LRRs, Serine-threonine kinase	Cytoplasm/trans-membrane	LRRs and kinases	129

[a]LRRs = leucine-rich repeats.

characterization of avirulence genes and their expressed products indicates that commonalities between the genes or proteins are the exception and, even when present, have not been enlightening. It is somewhat calming, at this point, to draw an analogy with the mammalian immune system. Plant resistance genes may be derived from a relatively small pool of molecular types analogous to immunoglobin genes; that is, a few common molecular structures, but with a variable recognition site. Avirulence genes, on the other hand, could produce substances analogous to antigens and, therefore, could direct the synthesis of a large number of potential elicitor-active compounds. The only criterion common to the elicitor compounds apparent at this time is that they must be available to interact with the plant recognition apparatus when the pathogen invades the plant tissue. As indicated in the three elicitor/receptor models (Fig. 3–2), an avirulence gene product can act directly as an elicitor (i.e., the elicitor is a protein) or indirectly (i.e., it controls the synthesis of an elicitor). Examples of both direct and indirect modes of action have been found. The *Avr9* gene, which was cloned from the fungal pathogen *C. fulvum,* encodes an extracellular peptide that, once processed, induces the HR in tomato plants carrying the resistance gene *Cf9.*[106,132] The product of the *avrD* gene, from the bacterial pathogen *P. syringae* pv. *tomato,* does not elicit the HR in soybean cultivars with the corresponding resistance gene.[18,51151] Instead, expression of *avrD* leads to the accumulation of low molecular weight acyl glycosides, called syringolides, in the culture supernate that are sufficient to trigger the HR.[18,133,134]

Avr4 and Avr9 from Cladosporium fulvum*: Gene Products that Act as Elicitors*

Race-specific proteinaceous elicitors accumulate in intercellular fluids during compatible interactions between races of *C. fulvum* with *Avr4* or *Avr9* activities and a susceptible cultivar of tomato (without the *Cf4* or *Cf9* genes for resistance).[73,74,75] A 28–amino-acid peptide purified from the fluids after inoculation with a strain containing *Avr9* caused the HR only on tomato cultivars with *Cf9*,[75] whereas a 106–amino-acid peptide from fluids after inoculation with a strain containing *Avr4* caused the HR only on tomato with *Cf4*.[64] The N-terminal amino acid sequences were determined for both avirulence genes, and oligonucleotide probes were designed that allowed the isolation of fungal clones from a cDNA library of a compatible *C. fulvum*-tomato interaction involving races carrying the functional *Avr4* and *Avr9* genes.[64,106] Confirmation that the two genes conferred avirulence to appropriate tomato cultivars was provided by introduction of the genes into virulent races of *C. fulvum* (a change from virulent to avirulent) and by gene disruption of strains containing the functional avirulence genes (a change from avirulent to virulent).[64,76,104] *Avr4* and *Avr9* genes encode proteins of a larger size than the peptide elicitors extracted from the intercellular fluids, and, in both cases, the proteins are translated as pre-pro-proteins. The pre-proteins are first cleaved upon extracellular targeting and then processed in the extracellular spaces into the mature elicitor by plant and/or fungal proteases.[64,76]

The available data do not yet provide a clear picture of how resistance is elicited in *Avr9-Cf9* interactions. Although the data are consistent with the avirulence gene product acting as an elicitor (Model 1), whether the host resistance gene product is the receptor is not yet known. In preliminary binding studies, iodinated AVR9 elicitor peptide bound to plasma membranes from leaves of both *Cf0* (susceptible control) and *Cf9* tomato genotypes, indicating that high-affinity binding sites exist in or on plasma membranes of both genotypes (unpublished results described in references 11 and 136). Such results raise the question of how the *Cf9* resistance gene and AVR9 receptor are related. Both *Cf9* and *Cf0* tomato genotypes contain members of a family of *Cf9*-homologues,[124] meaning that both lines could produce CF9–like proteins. The structure of the *Cf9* gene product is consistent with a glycoprotein with a cytoplasmic C-terminal domain and a large, extracellular domain containing LRRs.[124] LRR domains are thought to mediate protein-protein interactions. Thus, one of several models predicts that the AVR9 elicitor binds to the LRR domains of both CF9 and the homologue from *Cf0* tomato, but that the signal for resistance is transduced only by the CF9/AVR9 interaction because of differences in the cytoplasmic domains of the proteins.[11]

Other examples where avirulence is thought to be mediated by a proteinaceous elicitor also are provided by fungal pathogens. NIP1 is produced during *R. secalis* infection of barley and causes avirulence on cultivars carrying the *Rrs1* gene.[77,78] Many *Phytophthora* sp. produce small extracellular proteins (approx. 10 kD) called *elicitins* that induce necrosis on tobacco, radish, and turnip cultivars.[79–84] The *parA1* gene, encoding the elicitin parasiticein, from *Pseudomonas parasitica* was found to be a member of a complex multigene family.[82,83] Finally, the avirulence gene *PWL2,* which was cloned from a rice strain of *Magneporthe grisea,* encodes a 145–amino-acid, glycine-rich protein and confers avirulence to weeping lovegrass. The protein has a putative signal peptide, suggesting that it may function directly to confer avirulence.[16]

avrD *from* P. s. *pv.* tomato: *Synthesis of a Nonproteinaceous Elicitor*

P. syringae pv. *tomato,* a bacterial pathogen of tomato, elicits the HR on all cultivars of soybean and, therefore, is unable to cause disease in this plant. Kobayashi et al.[17] identified a single gene, called *avrD,* from *P. syringae* pv. *tomato* that, when transferred to *P. syringae* pv. *glycinea,* accounted for the HR on soybean cultivars with the *Rpg4* gene for resistance.[137] The 34 kd protein, which is encoded by *avrD,* is not exported from the bacterial cell, does not contain an apparent membrane spanning domain, and does not elicit the HR when infiltrated into soybean with *Rpg4*.[18] Thus, the *avrD* gene product likely does not play a direct role in race-specific elicitation of resistance.

Keen and coworkers[18] postulated that the *avrD* gene product is an enzyme that directs the synthesis of elicitor-active compound(s). Culture filtrates from bacteria (including *E. coli*) expressing the cloned gene contained a low molecular weight compound that elicited the HR only in cultivars of soybean carrying the complementary resistance gene *Rpg4*.[18] Culture filtrates from strains with a nonfunctional allele of *avrD* had little or no elicitor activity.[18] The elicitor-active compounds from the cultures are acyl glycosides called *syringolides 1* and *2*.[133,134] Based on the proposed synthetic scheme for the syringolides, *avrD* likely encodes a protein that catalyzes the condensation of xylulose with a ß-keto acid (either ß-hydroxydecanoic acid or ß-hydroxyoctanoic acid) to form the syringolides.[133–135] Syringolides 1 and 2 are intriguing molecules, not only because they elicit the plant HR in hosts with only the *Rpg4* gene, but also because they are similar in structure to signal molecules from other organisms. Midland et al.[133] and Smith et al.[134] pointed out similarities between butanolides produced by *Streptomyces* (which induce sporulation and antibiotic production);[138] the lactone autoinducers from *Vibrio* spp.,[139,140] *Agrobacterium tumefaciens*,[141] *Erwinia carotovora*[142] and *Pseudomonas aeruginosa*;[143] and endogenous jasmonic acid and related signal molecules of plants.[144] The analogies in the structures of these molecules and their roles as signal molecules indicate that their modes of action (what they interact with and how) might be similar.

CONCLUSIONS

The gene-for-gene model, which was founded on classical genetic analyses of host and pathogen populations, has remained valid, even under the scrutiny of modern molecular genetic analysis. The few resistance loci that have been cloned are single dominant genes, and the avirulence genes, for the most part, are single gene traits. As more genes are cloned and the biochemical details of the interactions are revealed, exceptions to the model likely will be described. Two recent discoveries, *avrE* from *P. syringae* pv. *tomato*[52,53] and *avrPphD* from *P. syringae* pv. *phaseolicola*,[59] are possible exceptions to the "single gene traits." In both of these cases, two gene products seem to be required for the avirulence phenotype to be manifested. In certain interactions between cotton and *X. campestris* pv. *malvacearum*, a single avirulence gene appears to interact with more than one host resistance gene, and conversely, more than one avirulence gene can interact with a single host resistance gene. De Feyter and coworkers[45] found that *avrB4* from *X. campestris* pv. *malvacearum* confers avirulence to cotton lines with either of two unlinked resistance gene loci, *B1* and *B4*. Furthermore, *avrB4* and two other avirulence genes (*avrb6* and *avrB102*) interact with a single cotton locus, *B1,* to elicit resistance. Such interactions, as these authors have suggested, may be described more accurately as "gene-for-genes" relationships.[45] In the particular case of *X. campestris* pv. *malvacearum* and cotton,[45] overlapping recognition specificities may be consequences of the structural similarities in elicitors. The avirulence genes from *X. campestris* pv. *malvacearum* that are implicated to have overlapping specificities are members of the *avrBs3* gene family,[45] and, as such, would encode similar proteins. By contrast, no such similarities have been discerned between the *avrB* gene from *P. syringae* pv. *glycinea* and the *avrRpm1* gene from *P. syringae* pv. *maculicola,* yet both apparently interact with a single resistance gene (*RPM1*) in *Arabidopsis.*[22,55]

Although more than one avirulence gene may indeed be recognized by a single resistance gene, what appears to be a single gene may, in fact, be multiple resistance genes at a single locus. Many resistance loci are complex, and others may be found to have genes that were undetected previously because of low recombination frequencies between them.[145] The recently cloned genes *Pto* and *Cf4* are members of small gene families and a number of the copies are located at the same locus.[123,124]

Mounting evidence implicates functions other than just host-cultivar specificity for avirulence genes. Several of the genes have been shown to be involved in aggressiveness of the pathogen to its host. The induction of the *P. syringae* avirulence genes in planta and their control by genes also involved in pathogenicity (the *hrp* genes) are consistent with an alternative function in interactions with plants or in stressful environments. Although some avirulence genes, such as *Avr9* from *C. fulvum* and *avrD* from *P. syringae* pv. *tomato,* do not appear to be essential to the reproduction of disease symptoms, we cannot conclude that the genes are dispensible. Many phases in the life and ecology of a pathogen undoubtedly are bypassed during laboratory and greenhouse studies. In an evolutionary sense, the majority of avirulence genes of a dispensable or gratuitous nature likely would be eliminated rapidly from the population, and those recognition factors involving functions of vitality or fitness would persist. In circumstances where the avirulence gene controls aggressiveness or other functions, complete gene elimination without compensation may not be possible. Genetic adaptations in the pathogen that eliminate avirulence activity on the predominate host genotypes may need to be concomitant with changes maintaining the essential function. The *avrBs3* gene family may well be a case in point. The high number of *avrBs3*-like genes in *X. oryzae* pv. *oryzae* may reflect the selective pressure on this organism to maintain aggressiveness while at the same time avoid avirulence. Rice is cultivated extensively under conditions that are highly favorable to bacterial blight, and the pathogen frequently is exposed to new genotypes as farmers, over the centuries, have sought to maintain or increase yield. The bacterial strains that survive may be those that have had a reservoir of *avrBs3*-like homologues that condition aggressiveness. If selection negates one avirulence gene, another copy or recombination among members of the gene family may provide a gene that lacks avirulence activity but retains aggressiveness.

In spite of the large number of avirulence genes that have been characterized, our understanding of the role of the products of these genes in the induction of resistance remains amazingly superficial. Many of the events in the plant that occur after recognition have been characterized,[4] but the molecular interactions that result in the activation of these events remain largely speculative. At present, *Avr9* from *C. fulvum* offers the best hope for determining if the AVR9 peptide interacts directly or indirectly with the *Cf9* gene product. Similarly, the interactions of fenthion with a *Pto* homologue may provide insight into avirulence gene-resistance gene interactions.[162] Rapid progress in understanding these interactions is likely to occur. Certainly, as in the case of toxins, knowing the host molecule to which the toxin binds goes a long way toward defining how the toxin affects the host.[146] Because of the commonalities in the resistance genes so far cloned from plants, it is tempting, but risky, to speculate that once we have deciphered a few of the molecular recognition events, models can be derived that encompass many situations. On the other hand, a greater effort is needed if we are to understand the biochemical nature of the elicited signals.

The identification of avirulence genes that confer resistance to nonhosts has practical implications. The resistance genes corresponding to the avirulence genes could be identified and cloned from the nonhost plants and introduced into important crop species. For this strategy to be effective, the targeted pathogen of the crop species must have the corresponding avirulence gene. Kearney and Staskawicz[99] found that functional copies of *avrBs2,* which interacts with the pepper resistance gene *Bs2* to result in stable and effective field resistance to the pathogen *X. campestris* pv. *vesicatoria,* were present in many other pathovars of *X. campestris,* such as pvs. *phaseoli, campestris,* and *citri.* If *Bs2* were cloned from pepper and introduced into the host plant species for these *X. campestris* pathovars (bean, cabbage, citrus, respectively), the gene might provide stable genetic resistance to these important diseases. The identification of avirulence genes that function in pathogens of ideal model plants, such as *Arabidopsis,* is especially appealing, because the sources of resistance might be identified more easily and manipulated for introduction into the less amenable crop species.[10,147] At the same time, engineering of resistance into plants will require careful consideration of the physiological effects of elicitation. Mutations that lead to constitutive expression of systemic acquired resistance in *Arabidopsis,* for example, have resulted in stunting of the plants.[163] Such negative effects may offset the advantage of enhanced resistance to pathogens.

The cloning of the first avirulence gene was reported in 1984,[13] and the cloning of the first resistance gene that involved race-specific recognition was reported almost 10 years later.[123] Over the next 10 years, the combined characterizations of avirulence and resistance genes and their products and the mechanisms by which they interact should be unraveled and may give us insights into new approaches for molecular cloning in crop improvement.

ACKNOWLEDGMENTS

We thank J. Dangl, P. De Wit, and J. Mansfield for providing preprints of manuscripts in progress. We also thank J. Chittoor, W. Zhu, S. Young, and S. Hulbert for helpful discussions and reviewing the manuscript. This manuscript is contribution # 95–411–B from the Kansas Agricultural Experiment Station.

REFERENCES

1. Flor, H. H. 1942. Inheritance of pathogenicity in *Melampsora lini*. *Phytopathology* **28:**365–391.
2. Ellingboe, A. H. 1976. Genetics of host-parasite interactions. In *Encyclopedia of Plant Pathology,* (NS), Physiological Plant Pathology, Vol. 4, ed. R. Heitefuss and P. H. Williams, pp. 761–778. Springer-Verlag, Heidelberg.
3. Dixon, R. A. and M. J. Harrison. 1990. Activation, structure, and organization of genes involved in microbial defense in plants. *Advances in Genetics* **28:**165–234.
4. Dixon, R. A., M. J. Harrison, and C. J. Lamb. 1994. Early events in the activation of plant defense responses. *Annu. Rev. Phytopathol.* **32:**479–501.
5. Kiraly, Z. 1980. Defenses triggered by the invader: Hypersensitivity. In *Plant Disease,* Vol. 4, eds. J. G. Horsfall and E. B. Cowling, pp. 201–224. Academic Press, New York.
6. Atkinson, M. M. 1993. Molecular mechanisms of pathogen recognition by plants. *Adv. Plant Pathol.* **10:**35–64.
7. Greenberg, J. T., A. Guo, D. F. Klessig, and F. M. Ausubel. 1994. Programmed cell death in plants: a pathogen-triggered response activated coordinately with multiple defense functions. *Cell* **77:**551–563.
8. Dietrich, R. A., T. P. Delaney, S. J. Uknes, E. R. Ward, J. A. Ryals, and J. L. Dangl. 1994. *Arabidopsis* mutants simulating disease resistance. *Cell* **77:**565–577.
9. Hammond-Kosack, K. E., K. Harrison, and J. D. G. Jones. 1994. Developmentally regulated cell death on expression of the fungal avirulence gene *Avr9* in tomato seedlings carrying the disease-resistance gene *Cf9*. *Proc. Nat. Acad. Sci. USA* **91:**10445–10449.
10. Dangl, J. L. 1994. The enigmatic avirulence genes of phytopathogenic bacteria. In *Current Topics in Microbiology and Immunology,* Vol. 192; Bacterial Pathogenesis of Plants and Animals, ed. J. L. Dangl, pp. 99–114. Springer-Verlag, Berlin.
11. De Wit, P. J. M. 1995. Fungal avirulence genes and plant resistance genes: unravelling the molecular basis of gene-for-gene interactions. *Adv. Bot. Res..* **21:**147–185.

12. Stakman, E. C. and F. J. Piemeisel. 1917. Biologic forms of *Puccinia gramminis* on cereals and grasses. *J. Agric. Res.* **10:**429–495.
13. Staskawicz, B., D. Dahlbeck, and N. T. Keen. 1984. Cloned avirulence gene of *Pseudomonas syringae* pv. *glycinea* determines race-specific incompatiblity on *Glycines max* (L.) Merr. *Proc. Nat. Acad. Science USA* **81:**6024–6028.
14. Whalen, M. C., R. E. Stall, and B. J. Staskawicz. 1988. Characterization of a gene from a tomato pathogen determining hypersensitive resistance in non-host species and genetic analysis of this resistance in bean. *Proc. Nat. Acad. Science USA* **85:**6743–6747.
15. Whalen, M. C., J. F. Wang, F. M. Carland, M. E. Heiskell, D. Dahlbeck, G. V. Minsavage, J. B. Jones, J. W. Scott, R. E. Stall, and B. J. Staskawicz. 1993. Avirulence gene *avrRxv* from *Xanthomonas campestris* pv. *vesicatoria* specifies resistance on tomato line Hawaii 7998. *Mol. Plant-Microbe Interact.* **6:**616–627.
16. Valent, B. and F. G. Chumley. 1994. Avirulence genes and mechanisms of genetic instability in the rice blast fungus. In *Rice Blast Diseases,* eds. R. S. Zeigler, S. A. Leong and P. S. Teng, pp. 111–134. CAB International, Oxon, U.K.
17. Kobayashi, D. Y., S. J. Tamaki, and N. T. Keen. 1989. Cloned avirulence genes from the tomato pathogen *Pseudomonas syringae* pv. *tomato* confer cultivar specificity on soybean. *Proc. Nat. Acad. Science USA* **86:**157–161.
18. Keen, N. T., S. Tamaki, D. Kobayashi, D. Gerhold, M. Stayton, H. Shen, S. Gold, J. Lorang, H. Thordal-Christensen, D. Dahlbeck, and B. J. Staskawicz. 1990. Bacteria expressing avirulence gene D produce a specific elicitor of the soybean hypersensitive reaction. *Mol. Plant-Microbe Interact.* **3:**112–121.
19. Ronald, P. C., J. M. Salmeron, F. M. Carland, and B. J. Staskawicz. 1992. The cloned avirulence gene *avrPto* induces disease resistance in tomato cultivars containing the *Pto* resistance gene. *J. Bacteriol.* **174:**1604–1611.
20. Lorang, J. M., H. Shen, D. Kobayashi, D. Cooksey, and N. T. Keen. 1994. *avrA* and *avrE* in *Pseudomonas syringae* pv. tomato PT23 play a role in virulence on tomato plants. *Mol. Plant-Microbe Interact.* **7:**508–515.
21. Vivian, A., G. Atherton, J. Bevan, I. Crute, L. Mur, and J. Taylor. 1989. Isolation and characterization of cloned DNA conferring specific avirulence in *Pseudomonas syringae* pv. *pisi* to pea (*Pisum sativum*) cultivars, which possess the resistance allele, *R2. Physiol. Mol. Plant Pathol.* **34:**335–344.
22. Dangl, J. L., C. Ritter, M. J. Gibbon, L. A. J. Mur, J. R. Wood, S. Goss, J. Mansfield, J. D. Taylor, and A. Vivian. 1992. Functional homologues of the *Arabidopsis RPM1* disease resistance gene in bean and pea. *Plant Cell* **4:**1359–1369.
23. Fillingham, A. J., J. Wood, J. R. Beven, I. R. Crute, J. W. Mansfield, J. D. Taylor, and A. Vivian. 1992. Avirulence genes from *Pseudomonas syringae* pathovars *phaseolicola* and *pisi* confer specificity towards both host and non-host species. *Physiol. Mol. Plant Pathol.* **40:**1–15.
24. Wood, J. R., A. Vivian, C. Jenner, J. W. Mansfield, and J. D. Taylor. 1994. Detection of a gene in pea controlling nonhost resistance to *Pseudomonas syringae* pv. *phaseolicola. Mol. Plant-Microbe Interact.* **7:** 534–537.

25. Swarup, S., Y. Yang, M. T. Kingsley, and D. W. Gabriel. 1992. A *Xanthomonas citri* pathogenicity gene, *pthA*, pleiotropically encodes gratuitous avirulence on nonhosts. *Mol. Plant-Microbe Interact.* **5:**204–213.
26. Carney, B. F. and T. P. Denny. 1990. A cloned avirulence gene from *Pseudomonas solanacearum* determines incompatibility on *Nicotiana tobacum* at the host species level. *J. Bacteriol.* **172:**4836–4843.
27. Ma, Q. S., M. -F. Chang, J. -L. Tang, J. -X. Feng, M. -J. Fan, B. Han, and T. Liu. 1988. Identification of DNA sequences involved in host specificity in the pathogenesis of *Pseudomonas solanacearum* strain T2005. *Mol. Plant-Microbe Interact.* **1:**169–174.
28. Mellano, V. J. and D. A. Cooksey. 1988. Development of host range mutants of *Xanthomonas campestris* pv. *translucens. Appl. Environ. Microbiol.* **54:**884–889.
29. Waney, V. R., M. T. Kingsley, and D. W. Gabriel. 1991. *Xanthomonas campestris* pv. *translucens* genes determining host-specific virulence and general virulence on cereals identified by Tn*5-gusA* insertion mutagenesis. *Mol. Plant-Microbe Interact.* **4:**623–627.
30. Niepold, F., D. Anderson and D. Mills. 1985 Cloning determinants of pathogenesis from *Pseudomonas syringae* pv. *syringae. Proc. Nat. Acad. Science USA* **82:**406–410.
31. Lindgren, P. B., R. C. Peet, and N. J. Panopoulos. 1986. Gene cluster of *Pseudomonas syringae* pv. "*phaseolicola*" controls pathogenicity on bean plants and hypersensitivity on nonhost plants. *J. Bacteriol.* **168:**512–522.
32. Willis, D. K., J. J. Rich, and E. M. Hrabak. 1991. *hrp* genes of phytopathogenic bacteria. *Mol. Plant-Microbe Interact.* **4:**132–138.
33. Salmond, G. P. C. and P. J. Reeves. 1994. Secretion of extracellular virulence factors by plant pathogenic bacteria. *Annu. Rev. Phytopathol.* **32:**181–200.
34. Wei, Z. -H. and S. V. Beer. 1992. *HrpN* of *Erwinia amylovora* functions in secretion of harpin and is a member of a new protein family. *J. Bacteriol.* **175:**7958–7967.
35. He, S. Y., H. -C Huang, and A. Collmer. 1993. *Pseudomonas syringae* pv. *syringae* $harpin_{Pss}$: a protein that is secreted via the *Hrp* pathway and elicits the hypersensitive response in plants. *Cell* **73:**1255–1266.
36. Arlat, M., V. Van Gijsegem, J. C. Huet, J. C. Pernollet, and C. A. Boucher. 1994. PopA1, a protein which induces a hypersensitivity-like response on specific *Petunia* genotypes, is secreted via the *Hrp* pathway of *Pseudomonas solanacearum. European Mol. Biol. Org. J.* **13:**543–553.
37. Yucel, I., D. Slaymaker, C. Boyd, J. Murillo, R. I. Buzzell, and N. T. Keen. 1994. Avirulence gene *avrPphC* from *Pseudomonas syringae* pv. *phaseolicola* 3121: A plasmid-borne homologue of *avrC* closely linked to an *avrD* allele. *Mol. Plant-Microbe Interact.* **7:**677–679.
38. Yucel, I., C. Boyd, Q. Debnam, and N. T. Keen. 1994. Two different classes of *avrD* alleles occur in pathovars of *Pseudomonas syringae. Mol. Plant-Microbe Interact.* **7:**131–139.
39. Tamaki, S., D. Dahlbeck, B. Staskawicz, and N. T. Keen. 1988. Characterization and expression of two cloned avirulence genes cloned from *Pseudomonas syringae* pv. *glycinea. J. Bacteriol.* **170:** 4846–4854.

40. Ronald, P. C. and B. J. Staskawicz. 1988. The avirulence gene *avrBs1* from *Xanthomonas campestris* pv. *vesicatoria* encodes a 50–kD protein. *Mol. Plant-Microbe Interact.* **1:**191–198.

41. Galyov, E. E., S. Hakansson, and H. Wolf-Watz. 1994. Characterization of the operon encoding the YpkA Ser/Thr protein kinase and the YopJ protein of *Yersinia pseudotuberculosis. J. Bacteriol.* **176:**4543–4548.

42. Bonas, U., R. E. Stall, and B. J. Staskawicz. 1989. Genetic and structural characterization of the avirulence gene *avrBs3* from *Xanthomonas campestris* pv. *vesicatoria. Mol. Gen. Genet.* **218:**127–136.

43. Bonas, U., S. Fenselau, T. Horns, C. Marie, B. Moussian, M. Pierre, K. Wengelnik, and G. van den Ackerveken. 1994. HRP and avirulence genes of *Xanthomonas campestris* pv. *vesicatoria* controlling the interaction with pepper and tomato. In *Advances in Molecular Genetics of Plant-Microbe Interactions,* Vol. 3, eds. M. J. Daniels, J. A. Downie and A. E. Osbourn, pp. 7–64. Kluwer Academic, Boston.

44. Canteros, B., G. Minsavage, U. Bonas, D. Pring, and R. Stall. 1991. A gene from *Xanthomonas campestris* pv. *vesicatoria* that determines avirulence in tomato is related to *avrBs3. Mol. Plant-Microbe Interact.* **4:**459–465.

45. De Feyter, R., Y. Yang, and D. W. Gabriel. 1993. Gene-for-gene interactions between cotton R genes and *Xanthomonas campestris* pv. *malvacearum avr* genes. *Mol. Plant-Microbe Interact.* **6:**225–237.

46. Hopkins, C. M., F. F. White, S. H. Choi, A. Guo, and J. E. Leach. 1992. Identification of a family of avirulence genes from *Xanthomonas oryzae* pv. *oryzae. Mol. Plant-Microbe Interact.* **5:**451–459.

47. Bonas, U., J. Conrads-Strauch, and I. Balbo. 1993. Resistance in tomato to *Xanthomonas campestris* pv. *vesicatoria* is determined by alleles of the pepper-specific avirulence gene *avrBs3. Mol. Gen. Genet.* **238:**261–269.

48. Yang, Y., R. De Feyeter, and D. W. Gabriel. 1994. Host-specific symptoms and increased release of *Xanthomonas citri* and *X. campestris* pv. *malvacearum* from leaves are determined by the 102–bp tandem repeats of *pthA* and *avrb6,* respectively. *Mol. Plant-Microbe Interact.* **7:**345–355.

49. Hopkins, C. M. 1993. Characterization of an avirulence gene from *Xanthomonas oryzae* pv. *oryzae.* Ph.D. dissertation, Kansas State University, Manhattan, KS.

50. Herbers, K., J. Conrads-Strauch, and U. Bonas. 1992. Race-specificity of plant resistance to bacterial spot disease determined by repetitive motifs in a bacterial avirulence protein. *Nature* **356:**172–174.

51. Kobayashi, D., S. J. Tamaki, and N. T. Keen. 1990. Molecular characterization of avirulence gene D from *Pseudomonas syringae* pv. tomato. *Mol. Plant-Microbe Interact.* **3:**94–102.

52. Lorang, J. M. and N. T. Keen. 1995. Characterization of *avrE* from *Pseudomonas syringae* pv. *tomato:* A *hrp*-linked avirulence locus consisting of at least two transcriptional units. *Mol. Plant-Microbe Interact.* **8:**49–57.

53. Keen, N., S. L. Midland, C. Boyd, I. Yucel, T. Tsurushima, J. Lorang, and J. J. Sims. 1994. Syringolide elicitors specified by avirulence gene D alleles in *Pseudomonas syringae.* In *Advances in Molecular Genetics of Plant-Microbe Interactions,* Vol. 3,

eds. M. J. Daniels, J. A. Downie, and A. E. Osbourn, pp. 41–48. Kluwer Academic, Boston.

54. Staskawicz, B., D. Dahlbeck, N. Keen, and C. Napoli. 1987. Molecular characterization of cloned avirulence genes from race 0 and race 1 of *Pseudomonas syringae* pv. *glycinea*. *J. Bacteriol.* **169:** 5789–5794.
55. Innes, R. W., A. F. Bent, B. N. Kunkel, S. R. Bisgrove, and B. J. Staskawicz. 1993. Molecular analysis of avirulence gene *avrRpt2* and identification of a putative regulatory sequence common to all known *Pseudomonas syringae* avirulence genes. *J. Bacteriol.* **175:**4859–4869.
56. Huynh, T., D. Dahlbeck, and B. J. Staskawicz. 1989. Bacterial blight of soybeans; regulation of a pathogen gene determining host cultivar specificity. *Science* **245:**1374–1377.
57. Salmeron, J. M. and B. J. Staskawicz. 1993. Molecular characterization and *hrp* dependence of the avirulence gene *avrPto* from *Pseudomonas syringae* pv. tomato. *Mol. Gen. Genet.* **239:**6–16.
58. Shen, H. and N. T. Keen. 1993. Characterization of the promoter of avirulence gene D from *Pseudomonas syringae* pv. *tomato*. *J. Bacteriol.* **175:**5916–5924.
59. Mansfield, J., C. Jenner, R. Hockenhull, M. A. Bennett, and R. Stewart. 1994. Characterization of *avrPphE,* a gene for cultivar-specific avirulence from *Pseudomonas syringae* pv. *phaseolicola* which is physically linked to *hrpY* a new *hrp* gene identified in the halo-blight bacterium. *Mol. Plant-Microbe Interact.* **7:**726–739.
60. Rahme, L. G., M. N. Mindrinos, and N. J. Panopoulos. 1992. The genetic and transcriptional organization of the *hrp* cluster of *Pseudomonas syringae* pathovar *phaseolicola*. *J. Bacteriol.* **173:**575–586.
61. Xiao, Y., Y. Lu, S. Heu, and S. W. Hutcheson. 1992. Organization and environmental regulation of the *Pseudomonas syringae* pv. syringae 61 *hrp* cluster. *J. Bacteriol.* **174:**1734–1741.
62. Grignon, C. and H. Sentenac. 1991. pH and ionic conditions in the apoplast. *Annu. Rev. Plant Physiol.* **42:**103–128.
63. Hancock, J. G. and O. C. Huisman. 1981. Nutrient movement in host-pathogen systems. *Annu. Rev. Phytopathol.* **19:**309–331.
64. Joosten, M. H., T. J. Cozijnsen, and P. J. De Wit. 1994. Host resistance to a fungal tomato pathogen lost by a single base-pair change in an avirulence gene. *Nature* **367:**384–386.
65. van den Ackerveken, G. F., R. M. Dunn, A. J. Cozijnsen, J. P. Vossen, H. W. Van den Broek, and P. J. De Wit. 1994. Nitrogen limitation induces expression of the avirulence gene *avr9* in the tomato pathogen *Cladosporium fulvum*. *Mol. Gen. Genet.* **243:**277–285.
66. Knoop, V., B. Staskawicz, and U. Bonas. 1991. Expression of the avirulence gene *avrBs3* from *Xanthomonas campestris* pv. *vesicatoria* is not under the control of *hrp* genes and is independent of plant factors. *J. Bacteriol.* **173:**7142–7150.
67. Young, S. A., F. F. White, C. M. Hopkins, and J. E. Leach. 1994. AVRXa10 protein is in the cytoplasm of *Xanthomonas oryzae* pv. *oryzae*. *Mol. Plant-Microbe Interact.* **7:**799–804.

68. Grimm, C., W. Aufsatz, and N. J. Panopoulos. 1995. The *hrpRS* locus of *Pseudomonas syringae* pv. *phaseolicola* constitutes a complex regulatory unit. *Mol. Microbiol.* **15:**155–165.

69. Grimm, C. G. and N. J. Panopoulos. 1989. The predicted protein product of a pathogenicity locus from *Pseudomonas syringae* pv. *phaseolicola* is homologueous to a highly conserved domain of several procaryotic regulatory proteins. *J. Bacteriol.* **171:**5031–5038.

70. Felley, R., L. G. Rahme, M. N. Mindrinos, R. D. Frederick, A. Pisi, and N. J. Panapoulos. 1991. Genes and signals controlling the *Pseudomonas syringae* pv. *phaseolicola*-plant interaction. In *Advances in Molecular Genetics of Plant-Microbe Interactions,* eds. H. Hennecke and D. P. S. Verma, pp. 45–52. Kluwer Academic, Dordrecht.

71. Jenner, C., E. Hitchin, J. Mansfield, K. Walters, P. Betteridge, D. Teverson, and J. Taylor. 1991. Gene-for-gene interactions between *Pseudomonas syringae* pv. *phaseolicola* and *Phaseolus. Mol. Plant-Microbe Interact.* **4:**553–562.

72. Xiao, Y., S. Heu, J. Yi, Y. Lu, and S. Hutcheson. 1994. Identification of a putative alternate sigma factor and characterization of a multicomponent regulatory cascade controlling the expression of *Pseudomonas syringae* pv. *syringae* Pss61 *hrp* and *hrmA* genes. *J. Bacteriol.* **176:**1025–1036.

73. De Wit, P. J. G. M. and G. Spikman. 1982. Evidence for the occurrence of race- and cultivar-specific elicitors of necrosis in intercellular fluids of compatible interactions between *Cladosporium vulvum* and tomato. *Physiol. Plant Pathol.* **21:**1–11.

74. De Wit, P. J. G. M., J. E. Hofman, G. C. M. Velthuis, and J. A. Kuc. 1985. Isolation and characterization of an elicitor of necrosis isolated from intercellular fluids of compatible interactions of *Cladosporium fulvum* (syn. *Fulvia fulva*) and tomato. *Plant Physiology* **77:**642–647.

75. Scholtens-Toma, I. M. J. and P. J. M. De Wit. 1988. Purification and primary structure of a necrosis-inducing peptide from the apoplastic fluids of tomato infected with *Cladosporium fulvum* (syn. *Fulvia fulva*). *Physiol. Mol. Plant Pathol.* **33:**59–67.

76. van den Ackerveken, G. F., J. A. Van Kan, and P. J. De Wit. 1992. Molecular analysis of the avirulence gene *Avr9* of the fungal tomato pathogen *Cladosporium fulvum* fully supports the gene-for-gene hypothesis. *Plant J.* **2:**359–366.

77. Hahn, M., S. Jungling, and W. Knogge. 1993. Cultivar-specific elicitation of barley defense reactions by the phytotoxic peptide NIP1 from *Rhynchosporium secalis. Mol. Plant-Microbe Interact.* **6:**745–754.

78. Knogge, W., A. Gierlich, H. Hermann, P. Wernert, and M. Rohe. 1994. Molecular identification and characterization of the *nip 1* gene, an avirulence gene from the barley pathogen *Rhynchosporium secalis.* In *Advances in Molecular Genetics of Plant-Microbe Interactions* Vol. 3, eds. M. J. Daniels, J. A. Downie, and A. Osbourn, pp. 207–214. Kluwer Academic, Boston.

79. Ricci, P., P. Bonnet, J. C. Huet, M. Sallantin, F. Beauvais-Cante, M. Bruneteau, V. Billard, G. Michel, and J. C. Pernollet. 1989. Structure and activity of proteins from pathogenic fungi *Phytophthora* eliciting necrosis and acquired resistance in tobacco. *European J. Biochem.* **183:**555–563.

80. Ricci, P., F. Trentin, P. Bonnet, P. Venard, F. Mouton-Perronnet, and M. Bruneteau. 1992. Differential production of parasiticein, an elicitor of necrosis and resistance in tobacco, by isolates of *Phytophthora parasitica. Plant Pathol.* **41:**298–307.

81. Pernollet J. C., M. Sallantin, M. Salle-Tourne, and J. C. Huet. 1993. Elicitin isoforms from seven *Phytophthora* species: comparison of their physico-chemical and toxic properties to tobacco and other plant species. *Physiol. Mol. Plant Pathol.* **41:**427–435.

82. Kamoun, S., M. Young, C. Glascock, and B. M. Tyler. 1993a. Extracellular protein elicitors from *Phytophthora:* host specificity and induction of resistance to fungal and bacterial phytopathogens. *Mol. Plant-Microbe Interact.* **6:**15–25.

83. Kamoun, S., K. M. Klucher, M. D. Coffey, and B. M. Tyler. 1993b. A gene encoding a host-specific elicitor protein of *Phytophthora parasitica. Mol. Plant-Microbe Interact.* **6:**573–581.

84. Kamoun, S., M. Young, H. Foster, M. D. Coffey, and B. Tyler. 1994. Potential role of elicitins in the interaction between *Phytophthora species* and tomato. *Appl. Environ. Microbiol.* **60:**1593–1598.

85. Brown, I., J. Mansfield, I. Irlam, J. Conrads-Strauch, and U. Bonas. 1993. Ultrastructure of interactions between *Xanthomonas campestris* pv. *vesicatoria* and pepper, including immunocytochemical localization of extracellular polysaccharides and the *avrBs3* protein. *Mol. Plant-Microbe Interact.* **6:**376–386.

86. Wagner, W., M. Kuhn, and W. Goebel. 1988. Active and inactive forms of hemolysin (HylA) from *Escherichia coli. J. Biol. Chem.* **369:**39–46.

87. Hughes, C., J. P. Issartel, K. Hardie, P. Stanley, E. Koronakis, and V. Koronakis. 1992. Activation of *Escherichia coli* prohemolysin to the membrane-targeted toxin by HlyC-directed ACP-dependence fatty acylation. *FEMS Microbiol. Immunol.* **5:**37–43.

88. Rosquist, R., K. E. Magnusson, and H. Wolf-Watz. 1994. Target cells contact triggers expression and polarized transfer of *Yersinia* YopE cytotoxin into mammalian cells. *EMBO J.* **13:**964–972.

89. Fenselau, S., I. Balbo, and U. Bonas. 1992. Determinants of pathogenicity in *Xanthomonas campestris* pv. *vesicatoria* are related to proteins involved in secretion in bacterial pathogens of animals. *Mol. Plant-Microbe Interact.* **5:**390–396.

90. Gough, C. L., S. Genin, C. Zischek, and C. Boucher. 1992. *hrp* genes of *Pseudomonas solanacearum* are homologous to pathogenicity determinants of animal pathogenic bacteria and are conserved among plant pathogenic bacteria. *Mol. Plant-Microbe Interact.* **5:**384–389.

91. Huang, H., R. Lin, C. Chang, A. Collmer, and W. Deng. 1995. The complete *hrp* gene cluster of *Pseudomonas syringae* pv. *syringae* 61 includes two blocks of genes required for harpin$_{Pss}$ secretion that are arranged colinearly with *Yersinia ysc* homologues. *Mol. Plant-Microbe Interact.* **8:**733–746.

92. Huang, H., S. W. Hutcheson, and A. Collmer. 1992. The *Pseudomonas syringae* pv. *syringae* 61 *hrpH* product: an envelope protein required for elicitation of the hypersensitive response in plants. *J. Bacteriol.* **174:**6878–6885.

93. Yang, Y. and D. Gabriel. 1995. *Xanthomonas* avirulence/pathogenicity gene family encodes functional plant nuclear targeting signals. *Mol. Plant-Microbe Interact.* **8**:627–631.

94. Johnson, P. F. and S. L. McKnight. 1989. Eukaryotic transcriptional regulatory proteins. *Annu. Rev. Biochem.* **58**:799–839.

95. Gabriel, D. W. 1989. The genetics of plant pathogen population structure and host-parasite specificity. In *Plant-Microbe Interactions: Molecular and Genetic Perspectives.* Vol. 3 ed. T. Kosuge and E. W. Nester, pp. 343–379. Macmillan, New York.

96. Keen, N. T. 1990. Gene-for-gene complimentary in plant-pathogen interaction. *Annu. Rev. Genet.* **24**:447–463.

97. Knorr, D. A. and W. O. Dawson. 1988. A point mutation in the tobacco mosaic virus capsid protein gene induces hypersensitivity in *Nicotiana sylvestris. Proc. Nat. Acad. Science USA* **85**:170–174.

98. Culver, J. N. and W. O. Dawson. 1989. Point mutations in the coat protein gene of tobacco mosaic virus induce hypersensitivity in *Nicotiana sylvestris. Mol. Plant-Microbe Interact.* **4**:209–213.

99. Kearney, B. and B. J. Staskawicz. 1990. Widespread distribution and fitness contribution of *Xanthomonas campestris* avirulence gene *avrBs2. Nature* **346**:385–386.

100. Minsavage, G. V., D. Dahlbeck, M. C. Whalen, B. Kearney, U. Bonas, B. J. Staskawicz, and R. E. Stall. 1990. Gene-for-gene relationships specifying disease resistance in *Xanthomonas campestris* pv. *vesicatoria*-pepper interactions. *Mol. Plant-Microbe Interact.* **3**:41–47.

101. Mazzola, M., J. E. Leach, R. Nelson, and F. F. White. 1994. Analysis of the interaction between *Xanthomonas oryzae* pv. *oryzae* and the rice cultivars IR24 and IRBB21. *Phytopathology* **84**:392–397.

102. Swarup, S., R. De Feyter, R. H. Brlansky and D. W. Gabriel. 1991. A pathogenicity locus from *Xanthomonas citri* enables strains from several pathovars of *X. campestris* to elicit cankerlike lesions on citrus. *Phytopathology* **81**:802–809.

103. Choi, S. 1993. Restriction-modification systems and genetic variability of *Xanthomonas oryzae* pv. *oryzae.* Ph.D. dissertation, Kansas State University, Manhattan, KS.

104. Marmeisse, R., G. F. J. M. van den Ackerveken, T. Goosen, P. J. G. M. De Wit, and H. W. J. van den Broek. 1993. Disruption of the avirulence gene *Avr9* in two races of the tomato pathogen *Cladosporium fulvum* causes virulence on tomato genotypes with the complementary resistance gene *Cf9. Mol. Plant-Microbe Interact.* **6**:412–417.

105. Bauer, D. W., A. J. Bogdanove, S. V. Beer, and A.Collmer. 1994. *Erwinia chrysanthemi hrp* genes and their involvement in soft rot pathogenesis and elicitation of the hypersensitive response. *Mol. Plant-Microbe Interact.* **7**:573–581.

106. van Kan, J. A. L., G. F. van den Ackerveken, and P. J. G. M. De Wit. 1991. Cloning and characterization of cDNA of avirulence gene *Avr9* of the fungal pathogen *Cla-*

dosporium fulvum, causal agent of tomato leaf mold. *Mol. Plant-Microbe Interact.* **4:**52–59.

107. Dahlbeck, D. and R. E. Stall. 1979. Mutations for change of race in cultures of *Xanthomonas vesicatoria. Phytopathology* **69:**634–636.

108. Kearney, B., P. C. Ronald, D. Dahlbeck, B. J. Staskawicz. 1988. Molecular basis for evasion of plant host defense in bacterial spot disease of pepper. *Nature* **332:**541–543.

109. Leung, H., R. J. Nelson, and J. E. Leach. 1993. Population structure of plant pathogenic fungi and bacteria. *Adv. Plant Pathol.* **10:** 157–205.

110. De Feyter, R. and D. W. Gabriel. 1991. At least six avirulence genes are clustered on a 90–kilobase plasmid in *Xanthomonas campestris* pv. *malvacearum. Mol. Plant-Microbe Interact.* **4:**423–432.

111. Leach, J. E., A. Guo, P. Reimers, S. -H. Choi, C. M. Hopkins. 1995. Physiology of resistant interactions between *Xanthomonas oryzae* pv. *oryzae* and rice. In *Molecular Mechanisms of Bacterial Virulence,* eds., C.I. Kado and J.H. Crosa, pp. 551–560. Kluwer Academic Publishers, Dordrecht.

112. Callow, J. A. 1977. Recognition, resistance and the role of plant lectins in host parasite interactions. *Adv. Bot. Res.* **4:**1–49.

113. Keen, N. T. 1982. Specific recognition in gene-for-gene host-parasite systems. *Adv. Plant Pathol.* **2:**35–82.

114. Gabriel, D. W., D. C. Loschke, and B. G. Rolfe. 1988. Gene-for-gene recognition: The ion channel defense model. In *Molecular Genetics of Plant-Microbe Interactions,* eds. R. Palacios and D. P. S. Verma, pp. 3–14. APS Press, St. Paul, MN.

115. Gabriel, D. W. and B. G. Rolfe. 1990. Working models of specific recognition in plant-microbe interactions. *Annu. Rev. Phytopathol.* **28:**365–391.

116. Albersheim, P. and A. J. Anderson-Prouty. 1975. Carbohydrates, proteins, cell-surfaces, and the biochemistry of pathogenesis. *Annu. Rev. Plant Physiol.* **26:**31–52.

117. Denarie, J., G. Truchet, and J. C. Prome. 1994. *Rhizobium* nodulation factors: synthesis and induction of plant responses. In *Advances in Molecular Genetics of Plant-Microbe Interactions,* eds., M. J. Daniels, J. A. Downie, and A. E. Osbourn, pp. 81–90. Kluwer Academic, Boston.

118. Hooykaas, P. J. J. and A. G. M. Beijersbergen. 1994. The virulence system of *Agrobacterium tumefaciens. Annu. Rev. Phytopathol.* **32:**157–179.

119. Denarie, J., F. Debelle, and C. Rosenberg. 1992. Signaling and host range variation in nodulation. *Annu. Rev. Microbiol.* **46:**497–531.

120. Fisher, R. F. and S. R. Long. 1992. *Rhizobium*-plant signal exchange. *Nature* **357:**655–660.

121. Denarie, J. and J. Cullimore. 1993. Lipo-oligosaccharide nodulation factors: a new class of signaling molecules mediating recognition and morphogenesis. *Cell* **74:**951–954.

122. Carlson, R. W., N. P. J. Price, and G. Stacey. 1994. The biosynthesis of rhizobial lipo-oligosaccharide nodulation signal molecules. *Mol. Plant-Microbe Interact.* **6:** 684–695.

123. Martin, G. B., S. H. Brommonschenkel, J. Chunwongse, A. Frary, M. W. Ganal, R. Spivey, T. Wu, E. D. Earle, S. D. Tanksley. 1993. Map-based cloning of a protein kinase gene conferring disease resistance in tomato. *Science* **262:** 1432–1435.

124. Jones, D. A., C. M. Thomas, K. E. Hammond-Kosack, P. J. Balint-Kurti, and J. D. G. Jones. 1994. Isolation of the tomato *Cf-9* gene for resistance to *Cladosporium fulvum* by transposon tagging. *Science* **266:**789–793.

125. Bent, A. F., B. N. Kunkel, D. Dahlbeck, K. L. Brown, R. Schmidt, J. Giraudat, J. Leung, B. J. Staskawicz. 1994. *RPS2* of *Arabidopsis thaliana:* A leucine-rich repeat class of plant disease resistance genes. *Science* **265:**1856–1860.

126. Mindrinos, M., F. Katagiri, G. -L. Yu, and F. M. Ausubel. 1994. The *A. thaliana* disease resistance gene *RPS2* encodes a protein containing a nucleotide-binding site and leucine-rich repeats. *Cell* **78**: 1089–1099.

127. Lawrence, G. J., J. G. Ellis, and E. J. Finnegan. 1994. Cloning a rust-resistance gene in flax. In *Advances in Molecular Genetics of Plant-Microbe Interactions,* eds., M. J. Daniels, J. A. Downie, and A. E. Osbourn, pp. 303–306. Kluwer Academic, Boston.

128. Whitham, S., S. P. Dinesh-Kumar, D. Choi, R. Hehl, C. Corr, and B. Baker. 1994. The product of the tobacco mosaic virus resistance gene *N*: Similarity to toll and the interleukin-1 receptor. *Cell* **78:**1101–1115.

129. Song, W.-Y., G.-L. Wang, L.-L. Chen, H.-S. Kim, L.-Y. Pi, T. Holsten, J. Gardner, B. Wang, W.-X. Zhai, L.-H. Zhu, C. Fauquet, and P. Ronald. 1995. A receptor kinase-like protein encoded by the rice disease resistance gene, *Xa21. Science* **270:**1804–1806.

130. Martin, G. B. 1995. Molecular cloning of plant disease resistance genes. In *Plant-Microbe Interactions,* Vol. 1, eds., G. Stacey and N. Keen, pp. 1–32. Chapman & Hall, New York.

131. Kobe, B. and J. Deisenhofer. 1994. The leucine-rich repeat: a versatile binding motif. *Trends in Biochem.* **19:**415–421.

132. van den Ackerveken, G. F., P. Vossen, and P. J. De Wit. 1993. The AVR9 race-specific elicitor of *Cladosporium fulvum* is processed by endogenous and plant proteases. *Plant Physiol.* **103:**91–96.

133. Midland, S. L., N. T. Keen, J. J. Sims, M. M. Midland, M. M. Stayton, V. Burton, M. J. Smith, E. P. Mazzola, K. J. Graham, and J. Clardy. 1993. The structure of syringolides 1 and 2, novel C-glycosidic elicitors from *Pseudomomas syringae* pv. *tomato. J. Am. Chem. Soc.* **58:**2940–2945.

134. Smith, M. J., E. P. Mazzola, J. J. Sims, S. L. Midland, N. T. Keen, V. Burton, and M. M. Stayton. 1993. The syringolides: Bacterial glycosyl lipids that trigger plant disease resistance. *Tetrahedron Lett.* **34:**223–226.

135. Yucel, I., S. L. Midland, J. J. Sims, and N. T. Keen. 1994B. Class I and Class II *avr*D alleles direct the production of different products in gram-negative bacteria. *Mol. Plant-Microbe Interact.* **7:**148–50.

136. Honée, G. F. J. M., H. W. J. van den Ackerveken, T. J. van den Broek, M. H. A. J. Cozijnsen, R. L. Joosten, M. Kooman-Gersmann, J. Vervoort, R. Vogelsang,

P. Vossen, J. P. Wubben, and P. J. G. M. De Wit. 1994. Molecular characterization of the interaction between the fungal pathogen *Cladosporium fulvum* and tomato. In *Advances in Molecular Genetics of Plant-Microbe Interactions* Vol. 3, eds. M. J. Daniels, J. A. Downie, and A. E. Osbourn, pp. 199–206. Kluwer Academic, Boston.

137. Keen, N. T. and R. I. Buzzell. 1991. New disease resistance genes in soybean against *Pseudomonas syringae* pv. *glycinea*: evidence that one of them interacts with a bacterial elicitor. *Theo. Appl. Genet.* **81:**133–138.

138. Beppu, T. 1992 Secondary metabolites as chemical signals for cellular differentiation. *Gene* **115:**159–165.

139. Bainton, N. J., B. W. Bycroft, S. R. Chhabra, P. Stead, L. Gledhill, P. J. Hill, C. E. D. Rees, M. K. Winson, G. P. C. Salmond, G. S. A. B. Steward, and P. Williams. 1992. A general role for the *lux* autoinducer in bacterial cell signalling: control of antibiotic biosynthesis in *Erwinia. Gene* **116:**87–91.

140. Cao, J. and E. A. Meighen. 1989. Purification and structural identification of an autoinducer for the luminescence system of *Vibrio harveyi. J. Biol. Chem.* **264:** 21670–21676.

141. Zhang, L., P. J. Murphy, A. Kerr, and M. E. Tate. 1993. *Agrobacterium* conjugation and gene regulation by N-acyl-L-homoserine lactones. *Nature* **362:** 446–448.

142. Flego, D., M. Pirhonen, R. Heikinheimo, M. B. Karlsson, T. K. Palva, and E. T. Palva. 1993. Global and specific control of exoenzyme production and virulence in *Erwinia carotovora. Proceedings Symposium Molecular Genetics of Plant-Microbe Interactions,* Rutgers Univ.

143. Passador, L., J. M. Cook, M. J. Gambello, L. Rust, and B. H. Iglewski. 1993. Expression of *Pseudomonas aeruginosa* virulence genes requires cell-to-cell communication. *Science* **260:**1127–1130.

144. Gundlach, H., M. J. Muller, T. M. Kutchan, and M. H. Zenk. 1992. Jasmonic acid is a signal transducer in elicitor-induced plant cell cultures. *Proc. Nat. Acad. Science USA* **89:**2389–2393.

145. Bennetzen, J. L. and S. H. Hulbert. 1992. Organization, instability and evolution of plant disease resistance genes. *Plant Mol. Biol.* **20:**575–580.

146. Walton, J. D. and D. G. Panaccione. 1993. Host-selective toxins and disease specificity: perspectives and progress. *Annu. Rev. Phytopathol.* **31:**275–303.

147. Dangl, J. L. 1993. The emergence of *Arabidopsis thaliana* as a model for plant-pathogen interactions. *Adv. Plant Pathol.* **10:**127–155.

148. Napoli, C. and B. Staskawicz. 1987. Molecular characterization and nucleic acid sequence of an avirulence gene from race 6 of *Pseudomonas syringae* pv. *glycinea. J. Bacteriol.* **169:**572–578.

149. Tamaki, S. J., D. Y. Kobayashi, and N. T. Keen. 1991. Sequence domains required for the activity of avirulence genes *avrB* and *avrC* from *Pseudomonas syringae* pv. *glycinea. J. Bacteriol.* **173:**301–307.

150. Wanner, L. A., S. Mittal, and K. R. Davis. 1993. Recognition of the avirulence gene *avrB* from *Pseudomonas syringae* pv. *glycinea* by *Arabidopsis thaliana. Mol. Plant-Microbe Interact.* **6:**582–591.

151. Kobayashi, D. Y., S. J. Tamaki, D. J. Trollinger, S. Gold, and N. T. Keen. 1990. A gene from *Pseudomonas syringae* pv. *glycinea* with homology to avirulence gene D from *P. s.* pv. *tomato* but devoid of the avirulence phenotype. *Mol. Plant-Microbe Interact.* **3:**103–111.

152. Yucel, I. and N. T. Keen. 1994. Amino acid residues required for the activity of *avrD* alleles. *Mol. Plant-Microbe Interact.* **7:**140–147.

153. Whalen, M. C., R. W. Innes, A. F. Bent, and B. J. Staskawicz. 1991. Identification of *Pseudomonas syringae* pathogens of *Arabidopsis* and a bacterial locus determining avirulence on both *Arabidopsis* and soybean. *Plant Cell* **3:**49–59.

154. Kunkel, B. N., A. F. Bent, D. Dahlbeck, R. W. Innes, and B. J. Staskawicz. 1993. *RPS2,* and *Arabidopsis* disease resistance locus specifying recognition of *Pseudomonas syringae* strains expressing the avirulence gene *avrRpt2. Plant Cell* **5:**865–875.

155. Dong, X., M. Mindrinos, K. R. Davis, and F. M. Ausubel. 1991. Induction of *Arabidopsis* defense genes by virulent and avirulent *Pseudomonas syringae* strains and by a cloned avirulence gene. *Plant Cell* **3:**61–72.

156. Debener, T., H. Lehnackers, M. Arnold, and J. L. Dangl. 1991. Identification and molecular mapping of a single *Arabidopsis thaliana* locus determining resistance to a phytopathogenic *Pseudomonas syringae* isolate. *Plant J.* **1:**289–302.

157. Bavage, A. D., A. Vivian, G. T. Atherton, J. D. Taylor, and A. N. Malik. 1991. Molecular genetics of *Pseudomonas syringae* pathovar *pisi:* plasmid involvement in cultivar-specific incompatibility. *J. Gen. Microbiol.* **137:**2231–2239.

158. Cournoyer, B., J. D. Sharp, A. Astuto, M. J. Gibbon, J. D. Taylor, and A. Vivian. 1995. Molecular characterization of *Pseudomonas syringae* pathovar *pisi* plasmid-borne avirulence gene *avrPpiB* which matches the *R3* resistance locus in pea. *Mol. Plant-Microbe Interact.* **8:**700–708.

159. Shintaku, M. H., D. A. Kluepfel, A. Yacoub, and S. S. Patil. 1989. Cloning and partial characterization of an avirulence gene from race 1 of *Pseudomonas syringae* pv. *phaseolicola. Physiol. Mol. Plant Pathol.* **35:**313–322.

160. Swanson, J., B. Kearney, D. Dahlbeck, and B. Staskawicz. 1988. Cloned avirulence gene of *Xanthomonas campestris* pv. *vesicatoria* complements spontaneous race-change mutants. *Mol. Plant-Microbe Interact.* **1:**5–9.

161. Parker, J. E., C. E. Barber, M. J. Fan, and M. J. Daniels. 1993. Interaction of *Xanthomonas campestris* with *Arabidopsis thaliana:* characterization of a gene from *X. c.* pv. *raphani* that confers avirulence to most *A. thaliana* accessions. *Mol. Plant-Microbe Interact.* **6:**216–224.

162. Martin, G. B., A. Frary, T. Wu, S. Brommonschenkel, J. Chunwongse, E. Earle, and S. Tanksley. 1994. A member of the tomato *Pto* gene family confers sensitivity to fenthion resulting in rapid cell death. *Plant Cell* **6:**1543–1552.

163. Bowling, S. A., A. Guo, H. Cao, A. S. Gordon, D. F. Klessig, and X. Dong. 1994. A mutation in *Arabidopsis* that leads to constitutive expression of systemic acquired resistance. *Plant Cell* **6:**1845–1857.

CHAPTER

4

Saponin Detoxification by Phytopathogenic Fungi

A. E. Osbourn, J. P. Wubben, and M. J. Daniels

INTRODUCTION

Saponins are an important group of glycosylated secondary metabolites that are widely distributed in the plant kingdom.[1] They are also found in some marine animals such as sea cucumbers and starfish.[2] Saponins are constituents of many plant drugs (e.g., digitoxin) and folk medicines (e.g., ginseng, ginger, and licorice) and are reported to have a wide range of effects on human health.[1,2,3,4,5,6] Some of these effects are detrimental, but because the majority are beneficial, there has been considerable interest in isolating and characterizing saponins and in testing their biological properties. In addition to the medicinal properties of saponins, other characteristics associated with these molecules include piscicidal, insecticidal, and molluscicidal activity; allelopathic action; antinutritional effects; and sweetness and bitterness, but this list is far from complete.[1,2,3,4,5] Many saponins have potent antifungal activity and, because they are often present in relatively high levels in healthy plants, they may contribute to the resistance of plants to attack by fungi.[3,6,7,8,9] Saponins may therefore be regarded as preformed antimicrobial agents. The term *phytoanticipin* has been proposed to distinguish preformed inhibitory substances of this nature from phytoalexins, which are synthesized in response to pathogen attack.[10] The role of saponins in disease resistance and the ability of some fungal pathogens to tolerate or detoxify the saponins of their host plants are the focus of this review. For more detailed information about the isolation, analysis, structures, and properties of saponins the reader should consult a recent comprehensive book by Hostettmann and Marston.[2]

WHAT ARE SAPONINS?

Saponins can be divided into three major groups depending on the structure of the aglycone (known as the *sapogenin*), which may be a steroidal glycoalkaloid, a steroid, or a triterpenoid.[1,2,3,4] These molecules are all derived from squalene but by different routes (Fig. 4–1). True triterpenoids contain 30 carbon atoms, while

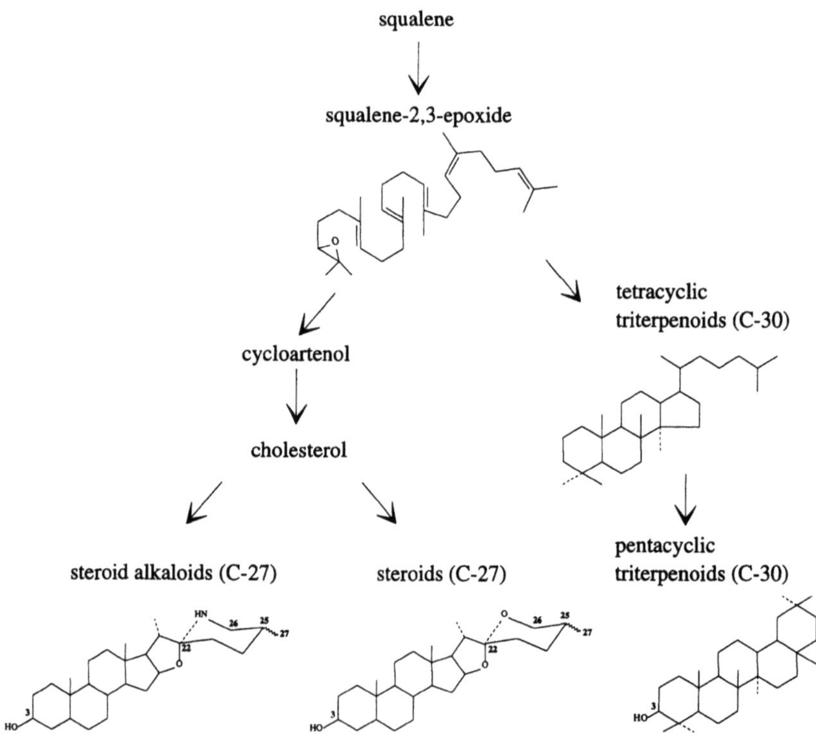

Figure 4–1. Saponin biosynthetic pathways in plants.

steroids (C 27) are synthesized from a C-30 intermediate by oxidative cleavage of three methyl groups. Steroidal alkaloids differ from steroids in that they contain a nitrogen atom, which renders the molecule basic. Although the enzymes involved in the cyclization of squalene and oxidosqualene have been studied in some detail,[11] in general the biosynthesis of the sapogenins is not well understood.

Sapogenins belonging to all three groups usually have a sugar chain attached at the hydroxyl group at the C-3 position via an ether linkage, and some also have sugar chains attached elsewhere on the molecule.[1,2,3,4] These sugar chains may be linear or branched, and usually consist of two to five monosaccharide residues. The addition of sugars to the C-3 position is catalyzed by specific transferases. Uridine diphosphate-dependent glycosyl transferases with specificities for the appropriate plant sapogenins have been characterized from a number of plants including oats,[12] tomato,[13] eggplant,[14,15,16] and potato.[17–19]

Saponins with a single sugar chain are referred to as *monodesmosidic.* Those with two sugar chains are known as *bisdesmosidic* and, in addition to having the sugar chain at C-3, they have a second sugar moiety attached at C-28 (triterpenoid saponins) or at C-26 (steroidal saponins). Bisdesmosidic saponins lack many of the properties and biological activities of monodesmodic saponins.[1,2,3,4] However, removal of the sugar at the C-26 or C-28 position gives the amphipathic monodesmosidic form, which is usually biologically active. For a number of bisdesmosidic saponins this conversion has been shown to occur in planta in response to tissue damage and is carried out by specific plant glycosyl hydrolases.[8,20,21,22]

ACTIVATION OF BISDESMOSIDIC SAPONINS BY PLANT GLYCOSYL HYDROLASES

Two plant enzymes that hydrolyze D-glucose from the C-26 position of bisdesmosidic saponins have been characterized in detail. The oat enzyme avenacosidase activates the foliar oat saponins avenacosides A and B by converting them into the biologically active monodesmosidic 26–desglucoavenacosides (26 DGAs) (Fig. 4–2A). This enzyme is specific for the C-26 glucose and does not remove sugars from the C-3 sugar chain.[20,21] A second enzyme with hydrolytic activity toward bisdesmosidic saponins has been purified from the herbaceous Asian plant *Costus speciosus.*[22] This plant contains the bisdesmosidic saponin protogracillin, which is a furostanol glycoside. Protogracillin is converted into the spirostanol glycoside gracillin by enzymatic cleavage of the C-26 glucose bond (Fig. 4–2B). The enzyme that carries out this conversion is known as furostanol glycoside 26–*O*—glucosidase (F26G) and is highly specific for furostanol glycosides. Both of these enzymes belong to the family 1 class of glycosyl hydrolases defined by Henrissat.[22,23,24] Other members of this family include the plant

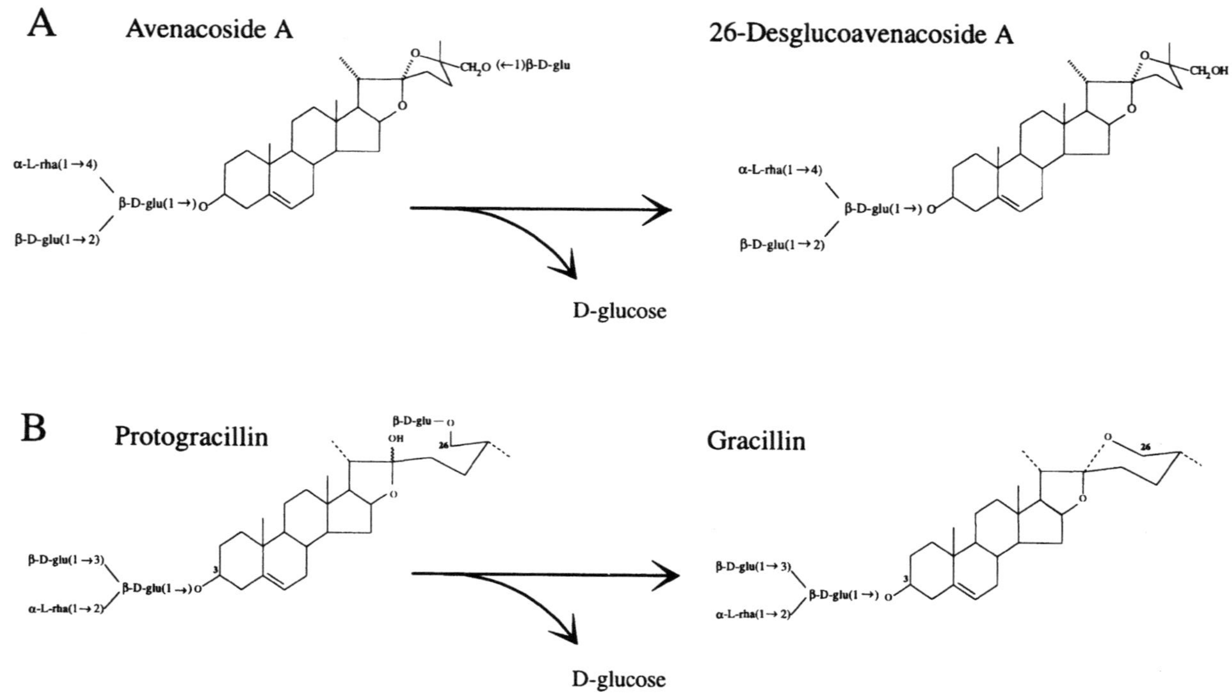

Figure 4–2 Activation of bisdesmosidic saponins by plant glucosyl hydrolases. (*A*) Conversion of the foliar oat saponin avenacoside A to 26–desglucoavenacoside A by avenacosidase. (*B*) Conversion of protogracillin to gracillin by the *Costus speciosus* enzyme F26G.

enzymes myrosinase and linamarase, which are also involved in the generation of antimicrobial compounds from preformed substances (glucosinolates and cyanogenic glycosides respectively).[25,26]

OCCURRENCE AND DISTRIBUTION OF SAPONINS

Triterpenoid saponins are found primarily in dicotyledonous plants, but also in some monocots. Over 750 triterpene glycosides with over 360 sapogenins have been identified, most of which have oleanane (β-amyrin) or dammarane skeletons.[2,27,28,29,30] Steroid saponins are less common than the triterpenoids, and occur mainly in monocots such as the *Liliaceae, Dioscoraceae* and the *Agavaceae.*[2,31] However dicots such as foxglove also produce steroid saponins (e.g., digitonin). Steroidal glycoalkaloids are found in over 350 plant species, primarily in members of the family *Solanaceae* (which includes the major crop plants potato and tomato), and also in the *Liliaceae.* Over 75 naturally occurring aglycone structures are known that are based on a C-27 cholestane skeleton with additional nitrogen-containing rings or groups.[2,6]

Saponin content and composition may vary markedly depending on the genetic background of the plant material and the tissue type. Other factors affecting saponin content include the age and physiologic state of the plant and environmental factors.[1,2,3,4,5] Oats, in addition to containing the steroidal bisdesmosidic avenacoside saponins also contain another saponin family, the triterpenoid avenacins.[32,33] The distribution of the two types of saponin within the plant are mutually exclusive, the avenacosides being located in the leaves and shoots,[34,35] and the avenacins in the root.[36,37] The occurrence of these saponins is restricted primarily to the genus *Avena,* although the avenacins are also known to occur in the closely related species *Arrhenatherum elatius.*[38,39] Searches for natural variation in saponin content in *Avena* species revealed that the C-genome oat species *A. clauda, A. pilosa,* and *A. ventricosa* lack avenacosides A and B,[21] whereas the A-genome diploid *Avena longiglumis* is the only *Avena* species so far identified which lacks the oat root saponin avenacin A-1.[40]

Glycoalkaloids are usually found in all organs of the plant, with the highest concentrations in regions of high metabolic activity.[2,5] The leaves of different *Lycopersicon* species show considerable variation in α-tomatine content, and variation is also observed between different accessions of the same species.[41,42,43] Genetic analysis of progeny from crosses of tomato material with low and high glycoalkaloid levels indicates that variation in α-tomatine content is controlled by the segregation of two codominant alleles at a single locus.[42] The green fruits of tomato are particularly rich in α-tomatine, with levels decreasing in ripening fruits.[5] An interesting variant of *Lycopersicon esculentum* var. *cerasiforme* originating from Peru has been found to have high levels of α-tomatine in the ripe fruits.[44] The genetic determination of this trait is recessive and probably monogenic. Possibly the variant is mutant for the allele coding for an α-tomatine degrading enzyme. Such an enzyme has been purified from tomato leaves,[45] but similar activities have not been characterized in the fruit.

A particularly detailed study of the genetic regulation of saponin composition has been carried out for soybean, which contains a complex mixture of different saponins.[46] An extensive collection of soybean cultivars were grouped on the basis of their saponin content. The inheritance of saponin types could be explained by different combinations of 5 genes controlling the utilization of soyasapogenol glycosides as substrates. Four of these genes were involved in different glycosylation steps and one in the acetylation of a glucose hydroxyl group. The function of the genes was variety specific and organ specific.

MECHANISM OF ANTIMICROBIAL ACTION

There is good evidence that the major mechanism of toxicity of saponins to fungi is due to their membraneolytic action.[1,3] Saponins have been demonstrated to complex with membrane sterols causing the formation of pores and hence loss of membrane integrity.[8,31,47] This property has been exploited in the use of red blood cell lysis tests as a means of detecting and quantifying saponins.[2] Oomycete fungi such as *Pythium* and *Phytophthora* species lack membrane sterols and are highly resistant to saponins.[48] However, if they are grown in medium supplemented with exogenous sterols they can incorporate these into their membranes, with an associated decrease in resistance to saponins.[49] In contrast, fungi that normally contain sterols in their membranes become more resistant to saponins when grown in the presence of inhibitors of sterol biosynthesis.[50] Mutants of *Fusarium solani* defective in sterol biosynthesis showed increased resistance to the tomato steroidal glycoalkaloid, α-tomatine. Significantly, these mutants were also pathogenic to the green fruits of tomato, which are particularly high in α-tomatine and are resistant to the wild-type fungus.[51] The mechanism of resistance of plant cells to saponins is unclear. Saponins may be compartmentalized in the vacuole or other organelles,[2,5,52] the membranes of which may have intrinsic resistance to lysis due to low or altered sterol composition.[47,53]

The way in which saponins interact with membrane sterols is not fully understood, although various models have been proposed.[1] The most detailed information concerning the ability of saponins to complex with membrane sterols comes from steroidal glycoalkaloids, which may not behave in the same way as the other classes of saponins because of their basicity. The steroidal glycoalkaloid α-tomatine forms a stable 1:1 complex with sterols that have a free 3,β-hydroxyl group.[54] The activity of a number of steroidal glycoalkaloids has been demonstrated to be pH dependent, with greater activity at higher pH where the steroidal glycoalkaloid is in its unprotonated form.[55,56] Little is known about the effects of pH on membraneolytic activity of triterpenoid and steroidal saponins. However those saponins with acidic or basic groups may also be expected to be more effective when uncharged, since this may facilitate incorporation into the membrane.

Because of the great diversity in the structures of saponins and the problems of purifying individual saponins in quantity, information about the relationship between structure and biological function of groups of closely related saponins is scant. In general a branched sugar chain comprised of 4 to 5 monosaccharides at C-3 seems to give higher antifungal activity,[2,57] although the nature of the aglycone is also likely to be important.[3]

Roddick and coworkers experimented using pairs of structurally related glycoalkaloids with common aglycones but different carbohydrate moieties[6,58] (Fig. 4–3). These pairs of saponins have been shown to exhibit synergistic activity with respect to inhibition of fungal growth and lysis of red blood cells and sterol-containing liposomes. Thus although individual steroidal glycoalkaloids may show little biological activity when tested separately, pronounced activity may be observed when pairs of compounds are mixed. This synergism has been shown to be particularly effective at ratios similar to those in which the pairs occur naturally,[56] and its significance should be considered when testing antifungal effects of saponins in vitro. The basis of the synergistic effect clearly involves some kind of complementation between carbohydrate moieties, but the mechanism is unknown.[6] Steroidal glycoalkaloids have other biological properties in addition to their ability to lyse membranes (e.g., inhibition of acetylcholinesterase activity) that have been ascribed to the basic properties of these molecules. Paired compounds do not show synergism when tested for acetylcholinesterase inhibition.[6]

SAPONIN DETOXIFICATION BY PLANT PATHOGENIC FUNGI

Whereas some fungi display intrinsic resistance to saponins by virtue of membrane composition, others produce extracellular enzymes that detoxify the saponins of their host plant.[8,9,59] Detailed information on saponin detoxification by plant pathogenic fungi is available only for pathogens of oat[60,61,62,63,64,65] and tomato[8,66,67,68,69,70,71] (Table 4–1), although the widespread distribution of saponins in the plant kingdom suggests that the ability to detoxify saponins may be a common attribute of successful pathogens.

The saponins featured in the remainder of this review are illustrated in Figure 4–4. The oat saponins consist of the triterpenoid monodesmosidic root saponins avenacin A-1, B-1, A-2, and B-2, and the steroidal bisdesmosidic saponins avenacosides A and B. Avenacins A-1 and B-1 are esterified with N-methyl anthanilic acid (and hence are autofluorescent under ultraviolet illumination), and A-2 and

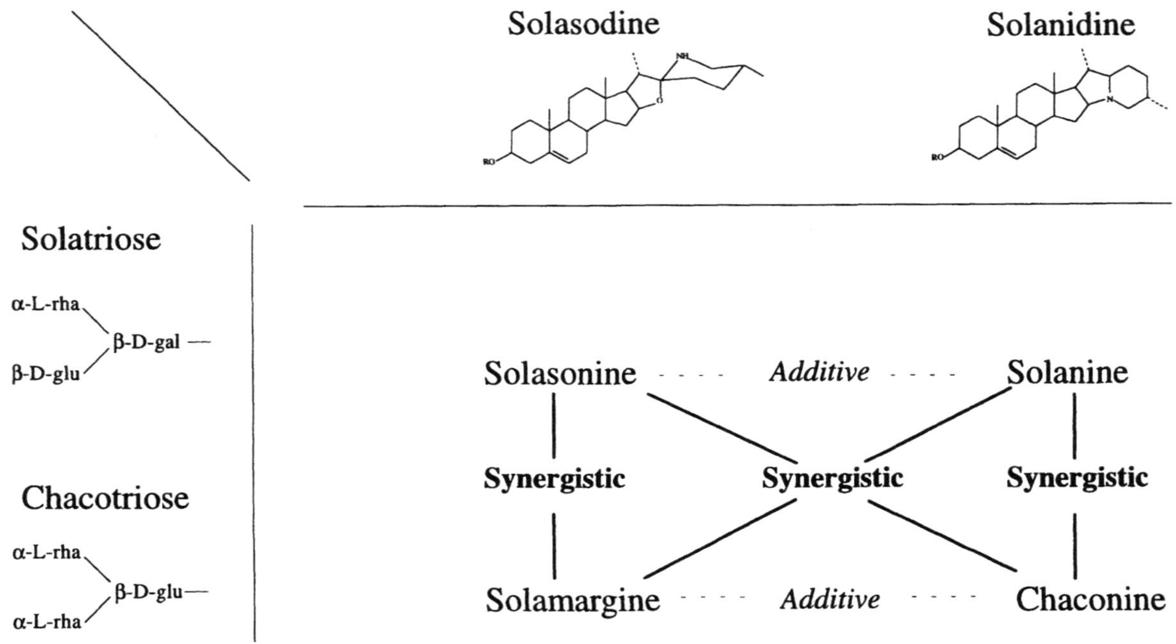

Figure 4–3 Synergism between glycoalkaloids with regard to inhibition of fungal growth, and lysis of red blood cells and liposomes. Pairs of glycoalkaloids with a common aglycone but different sugar chains display synergism (e.g., solasonine and solamargine, solanine, and chaconine). Molecules with the same sugar chain but different aglycones show only additive effects (e.g., solasonine and solanine, solamargine and chaconine). Synergism can also occur between glycoalkaloids with different aglycones (e.g. solasonine and chaconine, solamargine and solanine). Adapted from references 6 and 58.

Table 4–1 Enzymatic Detoxification of Oat and Tomato Saponins by Plant Pathogenic Fungi

Host Plant	Saponin	Pathogen	Reference
Oat	Avenacins	*Gaeumannomyces graminis* var. *avenae*	60,61,62,63
		Fusarium avenaceum	61,64
	26–DGAs	*Drechslera avenae*	65
		Septoria avenae	[a]
Tomato	α-tomatine	*Septoria lycopersici*	66,70,71
		Verticillium albo-atrum	67
		Botrytis cinerea	69
		Fusarium oxysporum f. sp. *lycopersici*	68
		Alternaria solani	8

[a] J. Wubben and A. Osbourn, unpublished.

B-2 are esterified with benzoic acid[32,33] (Fig. 4–4A). All four compounds have the same trisaccharide moiety attached to the C-3 carbon atom, consisting of β,1–2– and β,1–4–linked terminal D-glucose molecules attached via L-arabinose to the aglycone. The foliar oat saponins avenacosides A and B differ from each other only in that avenacoside B has an additional D-glucose molecule attached to the C-3 sugar chain[34,35] (Fig. 4–4B). As has already been mentioned, these two saponins are converted to the monodesmosidic forms 26 DGA A and B by the oat enzyme avenacosidase[20,21] (Fig. 4–2A). Tomato plants contain the steroidal glycoalkaloid α-tomatine, which is a monodesmosidic saponin with a tetrasaccharide group (known as *β-lycotetraose*) attached to the C-3 carbon[5] (Fig. 4–4C). β-lycotetraose consists of two molecules of D-glucose and one each of D-galactose and D-xylose. For all the saponins shown in Figure 4–4 the presence of an intact sugar chain at the C-3 position is essential for full toxicity[61,72] (J. Wubben and A. Osbourn, unpublished). The fungal enzymes that detoxify these molecules do so by hydrolysis of one or more sugars from this chain. Those fungi for which the mechanism of saponin detoxification has been well characterized are now considered in detail.

Detoxification of Oat Root Saponins by Gaeumannomyces graminis

Gaeumannomyces graminis var. *tritici* is one of the most damaging cereal root pathogens in temperate regions and causes the disease known as take-all.[73] Because the fungus is soilborne, fungicides have proved ineffective. Moreover,

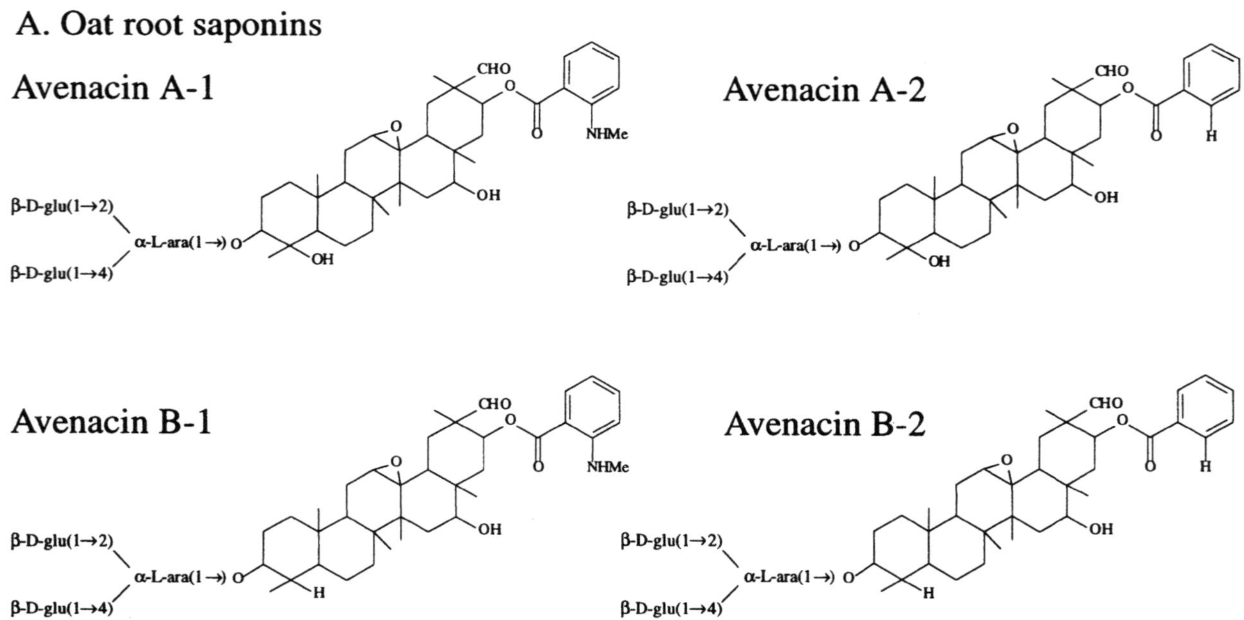

Figure 4–4 Structures of oat and tomato saponins. (*A*) The triterpenoid oat root saponins avenacins A-1, A-2, B-1, and B-2. (*B*) The bisdesmosidic steroidal foliar oat saponins, avenacosides A and B. The 26–desglucoavenacosides are formed by removal of the glucose molecules indicated by asterisks. (*C*) The tomato steroidal glycoalkaloid α-tomatine.

B. Oat leaf saponins

Avenacoside A

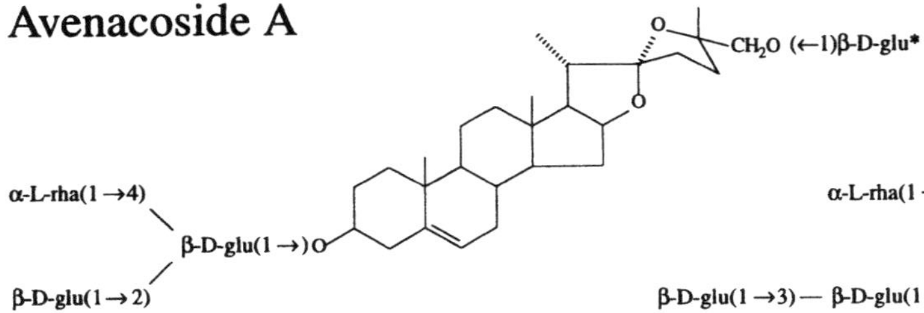

Avenacoside B

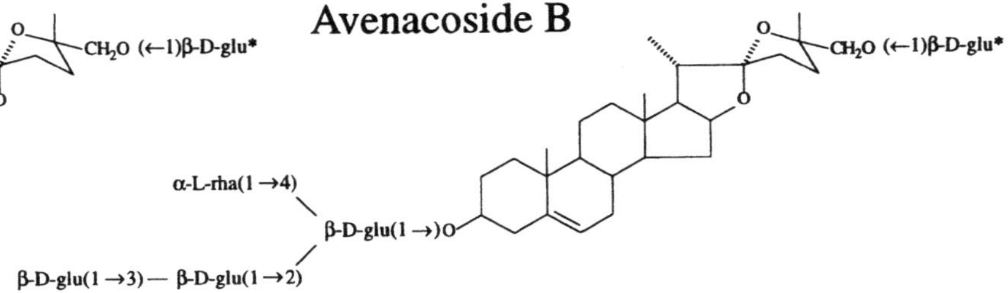

C. Tomato saponin

α-tomatine

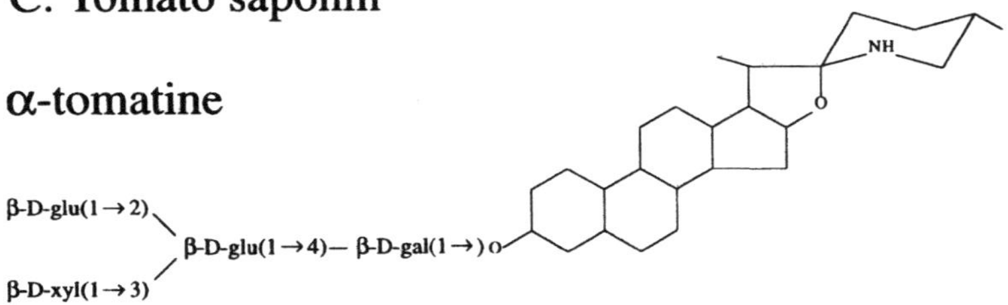

Figure 4–4 (*Con't.*)

searches for genetic resistance to take-all disease that could be incorporated into wheat have so far proved fruitless.[74] *G. graminis* var. *tritici* can infect wheat but is unable to infect oats. An oat-attacking form of *G. graminis, G. graminis* var. *avenae,*[75] can infect oats in addition to wheat, and is much less sensitive to the oat root avenacin saponins than is *G. graminis* var. *tritici*[60,61,62] (Table 4–2). This has led to speculation that the resistance of oats to *G. graminis* var. *tritici* may be attributable to the presence of avenacins in oat roots.[37,39,40,60,61,62,63]

Of the four avenacins, those that are esterified with N-methyl anthanilic acid (A-1 and B-1) are the most fungitoxic[38] (Table 4–3). Since avenacin A-1 is also the most abundant component of avenacin extracts[38] (Table 4–3), it is regarded as the one that is most likely to contribute to the resistance of oats to fungal attack.[38,40] Avenacin A-1 is known to occur exclusively in the root epidermis,[40] and so presents an effective barrier to attack by soil fungi. The observation that the one oat species that has been shown to lack avenacin A-1 (*A. longiglumis*) is susceptible to infection by *G. graminis* var. *tritici*[40] is consistent with the prediction that avenacin A-1 may determine resistance to this variety of *G. graminis.* A more stringent analysis of the importance of the avenacins in disease resistance involving the generation of chemically induced avenacin-minus mutants has been carried out. Preliminary analysis of these mutants indicates that they show increased susceptibility to both *G. graminis* var. *avenae* and var. *tritici* (R. Melton and A. Osbourn, unpublished).

G. graminis var. *avenae,* as well as another oat root-infecting fungus *Fusarium avenaceum,* are capable of enzymatic detoxification of the avenacins.[60,61,62] The detoxification of avenacin A-1 by *G. graminis* var. *avenae* has been studied in detail and involves the removal of β,1–2– and β,1–4–linked terminal D-glucose molecules to give the less toxic mono- and bis-deglucosylated forms.[61] This hydrolysis is carried out by an enzyme known as *avenacinase.*[60] All isolates of *G. graminis* var. *avenae* that have been tested are able to effectively deglucosylate avenacin A-1, whereas var. *tritici* isolates show little or no detoxifying activity.[60,61,62] The resistance of *G. graminis* var. *avenae* to avenacin A-1 is therefore correlated with the ability to actively detoxify the saponin, suggesting that avenacinase may be a determinant of host specificity. Genetic evidence has recently been presented that demonstrates that this is indeed the case. The evidence is as follows.

Table 4–2 Host Range and Avenacin Resistance of *Gaeumannomyces graminis* Varieties *tritici* and *avenae*

G. graminis Variety	Pathogenic to: Wheat	Oats	[avenacin A-1] Giving 100% Growth Inhibition (μg/ml)
tritici	+	–	5 to 10
avenae	+	+	50

Table 4–3 Autofluorescent Properties, Abundance, and Antifungal Activity of the Four Avenacins

Avenacin	Autofluorescence under UVL	% of Total Avenacins	% Growth Inhibition of *Gga* at 50 μM
A-1	+	47 to 60	81.8
B-1	+	5 to 7	75.1
A-2	-	30 to 43	53.4
B-2	-	3 to 6	65.3

Source: Adapted from reference 38.

Avenacinase was purified from culture filtrates of a pathogenic isolate of *G. graminis* var. *avenae* by successive fractionations involving free-flow isoelectric focusing and anion exchange and size exclusion high-performance liquid chromatography (HPLC).[62,63] The ability to remove both the β,1–2 and the β,1–4–linked D-glucose molecules was found to reside in a single protein of M_r 110 kD and with a pI of 4.6. Avenacinase gave equivalent rates of hydrolysis with avenacin A-1 and with the standard β-glucosyl hydrolase substrate *para*-nitrophenyl —D-glucopyranoside.[76] Poly A^+-RNA was extracted from a *G. graminis* var. *avenae* culture expressing avenacinase and used to construct a cDNA expression library in lambda gt11.[63] This library was screened with polyclonal antisera raised against purified avenacinase, and a number of cDNA clones containing overlapping DNA restriction fragments were isolated. These clones were in turn used to isolate a clone (pA3G1) bearing the avenacinase gene from a library of *G. graminis* var. *avenae* genomic DNA. The expression of pA3G1 in the heterologous fungus *Neurospora crassa* confirmed that this clone did indeed encode avenacinase. Southern blot analysis of *G. graminis* var. *avenae* genomic DNA digested with a range of restriction enzymes and using avenacinase cDNA as a probe indicated the existence of a single copy of the avenacinase gene. This gene was designated *AVN1*. An internal fragment of *AVN1* was used to construct a vector for transformation-mediated gene disruption (Fig. 4–5). The vector was introduced into *G. graminis* var. *avenae* using standard fungal transformation procedures, and a number of mutants were isolated in which the vector had inserted into the *AVN1* gene by homologous recombination. As expected, mutants disrupted in the *AVN1* gene were more sensitive to avenacin A-1 than the wild-type fungus, and lacked avenacinase activity. The mutants were tested for pathogenicity to wheat and oats. Symptoms produced on wheat were indistinguishable from those produced by the wild type. However, avenacinase-minus mutants did not give symptoms on oat seedlings (Fig. 4–6). These results indicate that avenacinase is required for pathogenicity to oats, but not for pathogenicity to wheat. Hence avenacinase is a determinant of host specificity, rather than a "basic compatibility factor" *sensu* Heath.[77]

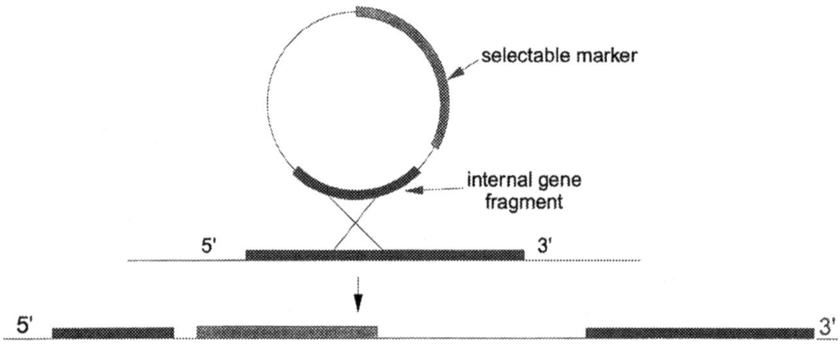

Figure 4–5 Targeted gene disruption in filamentous fungi.

Comparison of the predicted amino acid sequence of the *AVN1* gene product indicated that avenacinase is related to a subgroup of β-glucosyl hydrolases,[70] as expected from its biochemical function. These other β-glucosyl hydrolases are cellobiose-degrading enzymes from yeasts and *Trichoderma reesei,* and they belong to the glycosyl hydrolase family defined as family 3 by Henrissat.[78,79] The effects (if any) of these enzymes on saponins have not been described.

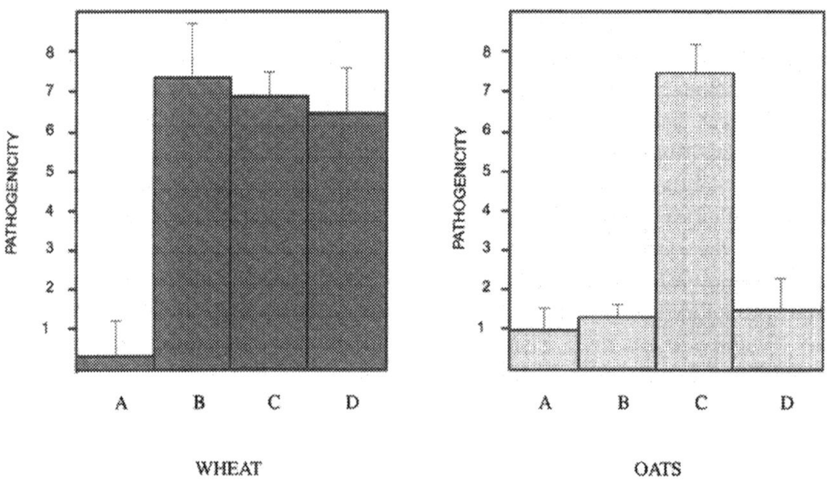

Figure 4–6 Pathogenicity of wild type and mutant *Gaeumannomyces graminis* strains to wheat and oats. (*A*) Mock inoculated control, (*B*) *G. graminis* var. *tritici,* (*C*) *G. graminis* var. *avenae,* (*D*) an avenacinase-minus mutant of *G. graminis* var. *avenae.* Pathogenicity was scored on an arbitrary scale of 0–8[63].

Other *Gaeumannomyces* Varieties and Species

G. graminis vars. *tritici* and *avenae* belong to a group of related fungi known as the *Gaeumannomyces/Phialophora* complex, which includes other *G. graminis* varieties, other *Gaeumannomyces* species, and a range of *Phialophora* species.[73] A detailed phylogenetic analysis of the relatedness of these fungi based on rDNA sequence comparisons has recently been completed.[80] Of these fungi, only *G. graminis* var. *avenae* shows significant resistance to avenacin A-1 and effectively degrades the saponin.[81] However *G. graminis* vars. *tritici* and *graminis,* as well as other *Gaeumannomyces* species such as *G. incrustans* and *G. cylindrosporus,* contain DNA sequences that hybridize to the *G. graminis* var. *avenae* *AVN1* gene.[59,82] Prolonged incubation of culture filtrates of these fungi with avenacin A-1 revealed that those fungi that contain the cross-hybridizing DNA sequences all have a weak ability to deglucosylate the saponin, while those that lacked cross-hybridizing DNA had no activity toward avenacin A-1. Moreover, the presence of enzyme activity was correlated with the occurrence of a protein of approximately 110 kD that cross-reacted with antiavenacinase antisera. These proteins have been named *avenacinase-like proteins* or ALPs. The *G. graminis* var. *tritici* ALP has been purified to homogeneity, thus establishing that the avenacin-deglucosylating activity is immunologically cross-reactive with avenacinase.[9] The specific activity of the *G. graminis* var. *tritici* ALP for avenacin A-1 was 25 times lower than that of avenacinase from *G. graminis* var. *avenae.*

The DNA that hybridizes to the *G. graminis* var. *avenae* *AVN1* gene has been cloned from *G. graminis* vars. *tritici* and *graminis.*[82] DNA sequence analysis indicates the existence of open reading frames (ORFs) that would encode proteins closely related to avenacinase (Table 4–4). Given that these ORFs appear to represent single copy genes in both *G. graminis* var. *tritici* and var. *graminis,* it is likely that the cloned DNA sequences encode the ALPs that have been biochemically characterized.

The function of the ALPs is unknown. Like avenacinase, these enzymes also have hydrolytic activity toward standard β-glucosyl hydrolase substrates, and it is possible that they may be required for nutrition during saprophytic growth.[9,81]

Table 4–4. DNA and Amino Acid Sequence Relatedness of ALPs to *G. graminis* var. *avenae* Avenacinase

ALP	% Nucleotide Identity	% Amino Acid Similarity	% Amino Acid Identity
var. *tritici*	95.9	92.9	95.7
var. *graminis*	88.5	84.7	90.9

However, they may also be important for detoxification of saponins other than the avenacins. *G. graminis* and other *Gaeumannomyces* species infect a wide range of grasses and may encounter saponins in these alternative host plants. Since little is known about the saponin content of wild grasses, the simplest way to test the hypothesis that ALPs may be important in determining host range would be to generate ALP-minus mutants by targeted gene disruption. These mutants could then be compared with the wild type for effects on pathogenicity to a diverse range of *Gramineae.*

DNA sequences that hybridize to the cloned *AVN1* gene are present in a range of other fungi, including both plant pathogenic fungi and fungi that do not infect plants.[70] These DNA sequences may encode β-glucosyl hydrolases required for saprophytic growth and with no role in saponin detoxification. Alternatively, they may encode enzymes required for saponin detoxification.

Detoxification of α-Tomatine by Fungal Pathogens of Tomato

A number of fungi that infect tomato produce enzymes known as *tomatinases,* which detoxify the steroidal glycoalkaloid saponin α-tomatine[8,66,67,68,69] (Fig. 4–7). The tomatinases of *Septoria lycopersici* and *Verticillium albo-atrum* hydrolyze a single terminal D-glucose molecule from β-lycotetraose.[66,67] *Alternaria solani* removes all four sugars to release the free monosaccharides and the agly-

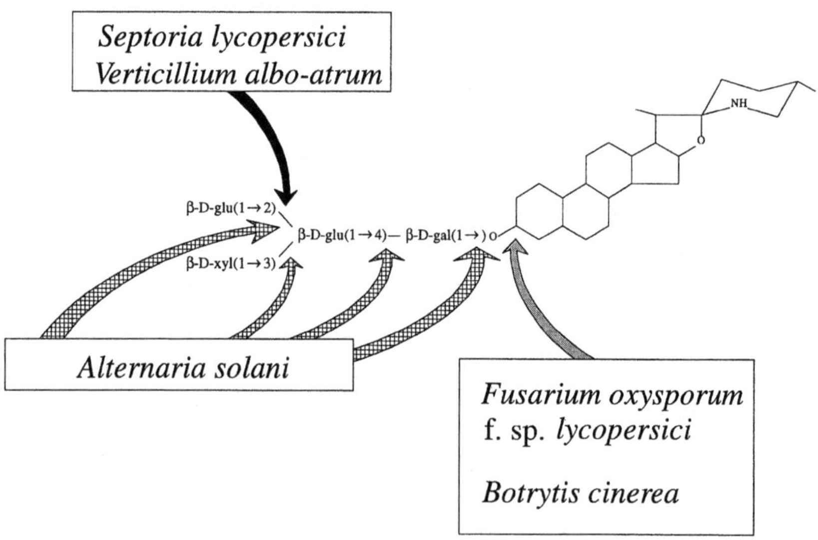

Figure 4–7 Detoxification of α-tomatine by fungal pathogens of tomato. The arrows indicate the bonds that are hydrolyzed by the various fungi.

cone, tomatidine,[8] whereas *Fusarium oxysporum* f. sp. *lycopersici* releases intact β-lycotetraose and tomatidine.[68] *Botrytis cinerea* is also reported to hydrolyze lycotetraose from α-tomatine as shown in Figure 4–7,[69] although this fungus also produces other enzymes that modify α-tomatine. These modifications involve the release of the terminal D-xylose (T. Quidde and P. Tudzynski, pers. comm.) and also the conversion of α-tomatine to an unidentified less polar compound that still contains the four sugar units.[82]

The mechanism of action of the tomatinase of *S. lycopersici* is similar to that of avenacinase, since both enzymes hydrolyze β,1–2–linked D-glucose residues from their saponin substrates.[61,62,66,70,71] Purification of the tomatinase activity from *S. lycopersici* revealed that this activity resided in a single protein that had similar properties to avenacinase and was recognized by the polyclonal antiavenacinase antisera.[70,71] Although avenacinase and tomatinase are related in terms of their physicochemical and immunological properties, there are clear differences in their substrate specificities that are in accordance with the host preferences of the two fungi[70] (Table 4–5). In parallel experiments, the *G. graminis* var. *avenae* avenacinase gene *AVN1* was used as a hybridization probe to screen a library of *S. lycopersici* cDNA.[70] A number of positive cDNA clones that cross-hybridized to each other and that contained common restriction fragments were isolated. DNA sequence analysis of a full-length cDNA clone revealed an open reading frame encoding a predicted protein clearly related to avenacinase (68% similar amino acids). Several lines of evidence indicated that this cDNA clone encoded tomatinase. First, internal amino acid sequences determined experimentally from the purified tomatinase protein could be identified within the deduced amino acid sequence.[70] Second, the cloned cDNA corresponded to a single cognate gene in the *S. lycopersici* genome.[70] Finally, this cDNA directed the synthesis of tomatinase in *Neurospora crassa* when expressed under the control of an appropriate promoter (Fig. 4–8) (R. Melton and A. Osbourn, unpublished). The generation of tomatinase-minus mutants of *S. lycopersici* by targeted gene disruption will allow the contribution of tomatinase to the *S. lycopersici* infection process to be tested.

Table 4–5. Substrate Specificity of *G. graminis* var. *avenae* Avenacinase and *S. lycopersici* Tomatinase

	Specific Activity[a]		
	avenacin A-1	α-tomatine	*p*NPG
Avenacinase	132.7	2.55	134.3
Tomatinase	0.002	8492.1	36.8

[a]nMoles glucose released/min/mg protein.

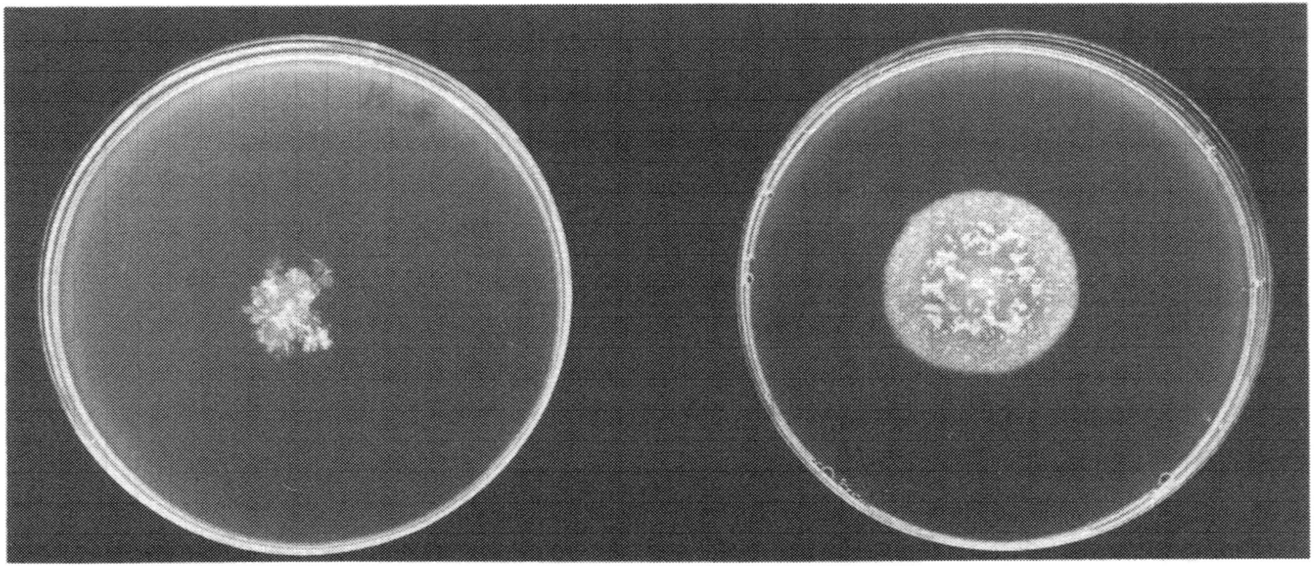

Figure 4–8 Expression of tomatinase in the heterologous fungus *Neurospora crassa*. The colony on the right is expressing *Septoria lycopersici* tomatinase: the colony on the left is a wild-type control. The colonies are growing on medium containing 250 μM α-tomatine. From R. Melton and A. Osbourn, unpublished.

Detoxification of the Foliar Oat Saponins, Avenacosides A and B

Lüning and Schlösser measured electrolyte leakage from the mycelium of a range of fungi in response to treatment with the antifungal monodesmosidic saponin 26 DGA B,[65] which is released from the bisdesmosidic avenacoside B in oat leaf tissue in response to damage. They found that foliar oat pathogens such as *Drechslera avenae* and *Septoria avenae* were insensitive to the membraneolytic action of the saponin, while *G. graminis* var. *avenae* and other oat root-infecting fungi were sensitive. The majority of nonpathogens of oat were also sensitive. Lüning and Schlösser also demonstrated that *D. avenae* was able to hydrolyze 26 DGAs to the aglycone nuatigenin by successive removal of sugars,[65] although the enzyme(s) responsible for this activity have not been purified.

Our studies of the significance of saponin detoxification in pathogenic fungi have recently been extended to include *Septoria avenae* (J. Wubben and A. Osbourn, unpublished). Isolates of this fungus can be subdivided into two *formae speciales*, namely f. sp. *avenae,* which infects oats but not wheat, and f. sp. *triticea,* which infects wheat but not oats. In a detached leaf assay we determined the pathogenicity of a number of isolates of *Septoria* to oat and wheat (Fig. 4–9). A positive correlation was found between pathogenicity to oats and ability to grow on artificial medium supplemented with extracts of oat leaves (Fig. 4–10). Thin layer chromatography of oat leaf extracts followed by bioassay tests revealed that the only antifungal substances detectable in these assays corresponded to 26 DGA A and B. In addition, it was shown that culture filtrates of oat-infecting isolates of *Septoria* contained an enzyme activity that was able to hydrolyze 26 DGAs A and B to compounds with little or no antifungal activity. The detoxification of the 26 DGAs by fungal enzymes involved the successive removal of the different sugar molecules at carbon 3. A single protein of 110 kD with a pI of 4.1 was purified from an oat-attacking isolate that was able to completely hydrolyze both 26 DGAs A and B into the aglycone, nuatigenin. The properties of this enzyme in terms of its molecular weight and pI are not dissimilar to those of avenacinase from *G. graminis* var. *avenae.*[62,63] However avenacosidase of *S. avenae* could not hydrolyze the oat root saponin avenacin A-1. It is clear from Southern blot analysis that isolates belonging to both f. sp. *avenae* and f. sp. *triticea* contain DNA sequences that hybridize to the *G. graminis* var. *avenae* avenacinase gene, although it is not yet known whether these sequences may encode the avenacosidase activity found in isolates of *S. avenae* f. sp. *avenae*. Determination of amino acid sequence of the purified fungal avenacosidase enzyme should permit a reverse genetic approach to be taken to isolate the cognate gene and to assess the degree of relatedness between avenacosidase and avenacinase. A DNA transformation procedure for *S. avenae* has been developed (J. Wubben and A. Osbourn, unpublished), so it should be feasible to address the role of avenacosidase in determining pathogenicity to oat following the approach taken for the *G. graminis* var. *avenae* avenacinase gene.

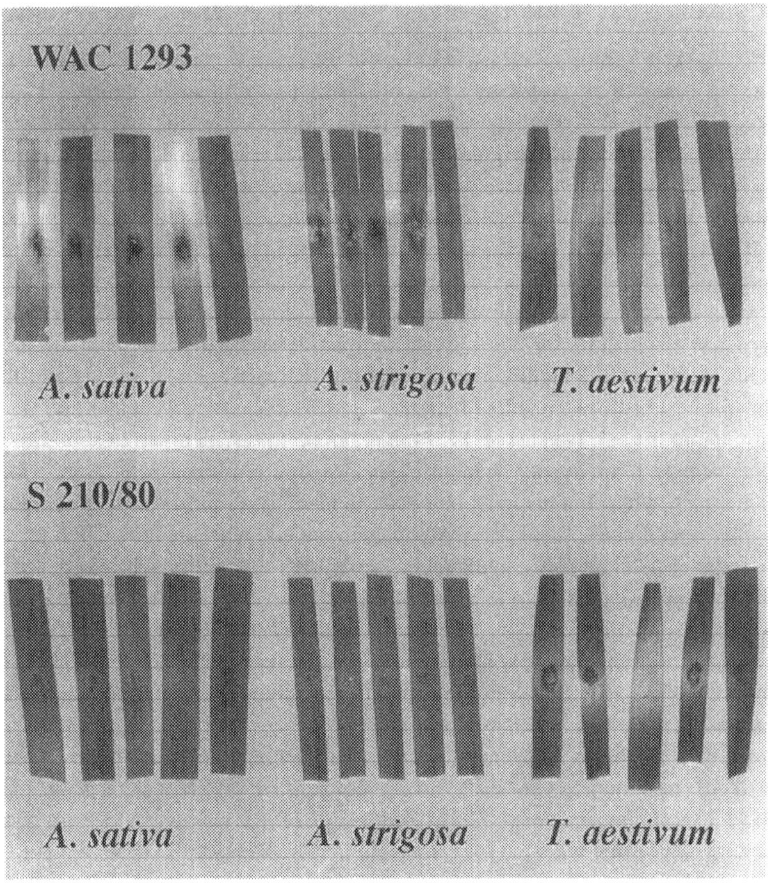

Figure 4–9. Detached leaf tests showing adaptation of two isolates of *Septoria avenae* (WAC 1293 and S210/80) to oats (*Avena sativa* and *A. strigosa*) (top panel) or wheat (*Triticum aestivum*) (bottom panel) respectively.

Saponin Detoxification and Fungal Pathogenicity—A General Phenomenon?

This review has summarized the current information about enzymatic detoxification of saponins by oat- and tomato-infecting fungi. For *G. graminis* var. *avenae* there is genetic evidence that avenacinase is a determinant of host range for oats.[63] The recent isolation of the DNA encoding tomatinase from *S. lycopersici*[70,71] should now allow the significance of saponin detoxification in pathogenicity to be tested for the interaction between *S. lycopersici* and tomato, assuming that targeted gene disruption can be achieved for this fungus. The application

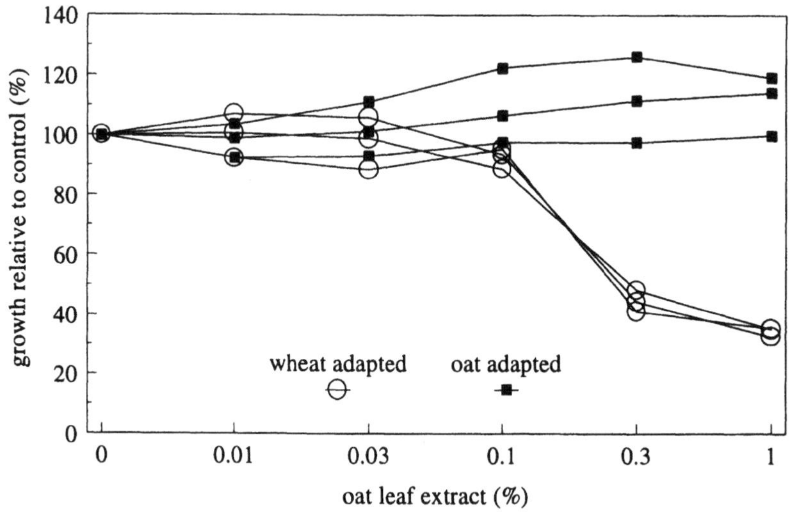

Figure 4–10. Effects of increasing amounts of oat leaf extract on the growth of wheat- and oat-adapted forms of *S. avenae*.

of similar approaches to other tomato pathogens and to *S. avenae* will extend these studies of saponin detoxification further.

DNA sequences related to the cloned avenacinase gene are found in *G. graminis* vars. *tritici* and *graminis,* and in a wide range of other phytopathogenic fungi that are not known to detoxify saponins. The significance of this observation is unclear. These sequences may encode β-glucosyl hydrolases required simply for fungal nutrition; alternatively, as in the case of *S. lycopersici,* they may encode saponin-detoxifying enzymes. Thus the cloned avenacinase and tomatinase genes or polymerase chain reaction (PCR) primers derived from their DNA sequences may allow the isolation of genes encoding similar enzymes from other phytopathogenic fungi. For many of these fungi the significance of saponin detoxification in the disease process may not have been studied. However, given the widespread distribution of saponins in the plant kingdom, it is quite likely that saponins may be important in determining the outcome of interactions between fungi that infect plants other than oats or tomato.

ACKNOWLEDGMENTS

The Sainsbury Laboratory is supported by the Gatsby Charitable Foundation. Jos Wubben is supported by the European Union Human Capital and Mobility Network grant, CHRX-CT93–0244.

We thank those who have shared their unpublished information with us.

REFERENCES

1. Price, K. R., I. T. Johnson and G. R. Fenwick. 1987. The chemistry and biological significance of saponins in food and feedingstuffs. *CRC Crit. Rev. Food Sci. Nutr.* **26:**27–133.
2. Hostettmann, K. A. and A. Marston. 1995. Saponins. *Chemistry and Pharmacology of Natural Products.* Cambridge University Press, Cambridge, UK.
3. Fenwick, G. R., K. R. Price, C. Tsukamota, and K. Okubo. 1992. Saponins. In *Toxic Substances in Crop Plants,* eds. J. P. D'Mello, C. M. Duffus and J. H. Duffus, pp. 284–327. The Royal Society of Chemistry, Cambridge.
4. Hostettmann, K., M. Hostettmann, and A. Marston. 1991. Saponins, *Methods Plant Biochem.* **7:**434–471.
5. Roddick, J. G. 1974. The steroidal glycoalkaloid α-tomatine. *Phytochemistry* **13:**9–25.
6. Roddick, J. G. Steroidal glycoalkaloids: Nature and consequences of bioactivity. In *Saponins: Chemistry and Biological Activity,* eds. G. Waller and K. Yamasaki, in press, Plenum, New York.
7. Tschesche, R. 1971. Advances in the chemistry of antibiotic substances from higher plants. In *Pharmacognosy and Phytochemistry,* eds. H. Wagner and L. Hörhammer, pp. 274–289. Springer-Verlag, Berlin.
8. Schönbeck, F. and Schlösser, E. 1976. Preformed substances as potential protectants. In *Physiological Plant Pathology,* eds. R. Heitefuss and P. H. Williams, pp. 653–678. Springer Verlag, Berlin.
9. Osbourn, A. E., P. Bowyer, P. Lunness, B. R. Clarke, and M.J. Daniels. 1994. Detoxification of plant saponins by fungi. In *Molecular Genetics of Plant-Microbe Interactions,* eds. M. J. Daniels, J. A. Downie and A. E. Osbourn, pp. 214–221. Kluwer, Dordrecht.
10. VanEtten, H. D., J. W. Mansfield, J. A. Bailey and E. E. Farmer. 1994. Two classes of plant antibiotics: Phytoalexins versus "Phytoanticipins." *Plant Cell* **9:**1191–1192.
11. Abe, I., M. Rohmer, and G. D. Prestwich. 1993. Enzymatic cyclization of squalene and oxidosqualene to sterols and triterpenes. *Chem. Rev.* **93:**2189–2206.
12. Kalinowska, M. and Z. A. Wojchiechowski. 1988. Substrate specificity of partially purified UDP-glucose:nuatigenin glucosyltransferase from oat leaves. *Plant Sci.* **55:**239–245.
13. Zimowski, J. 1995. Enzymatic glycosylation of tomatidine in tomato plants. In Abstracts (part 1), 210th American Chemical Society National Meeting (Division of Agriculture and Food Chemistry), August 20–24 1995, Chicago, 184. American Chemical Society, Washington D.C.
14. Paczkowski, C. and Z. A. Wojchiechowski. 1994. Glucosylation and galactosylation of diosgenin and solasodine by soluble glycosyltransferase(s) from *Solanum melongena* leaves. *Phytochemistry* **35:**1429–1434.
15. Wojciechowski, Z. A., C. Paczkowski, M. Kalinowska, and R. Woldański. 1995. Evidence for separate UDP-glucose:spirostanol and UDP-glucose: solasodine glucosyl transferases in *Solanum melongena.* In Abstracts (part 1), 210th Ameri-

can Chemical Society National Meeting (Division of Agriculture and Food Chemistry), August 20–24 1995, Chicago, 181. American Chemical Society, Washington D.C.

16. Wojciechowski, Z. A. and C. Paczkowski. 1995. Purification and properties of UDP-glucose:diosgenin/tiogogenin glucosyltransferase from *Solanum melongena* leaves. In Abstracts (part 1), 210th American Chemical Society National Meeting (Division of Agriculture and Food Chemistry), August 20–24 1995, Chicago, 182. American Chemical Society, Washington D.C.
17. Stapleton, A., P. V. Allen, M. Friedman, and W. R. Belknap. 1991. Purification and characterization of solanidine glucosyltransferase form the potato (*Solanum tuberosum*). *J. Agric. Food Chem.* **39:**1187–1193.
18. Zimowski, J. 1991. Occurrence of a glucosyltransferase specific for solanidine in potato plants. *Phytochemistry* **30:**1827–1831.
19. Bergenstråhle, A., E. Tillberg, and L. Jonsson. 1992. Characterization of UDP-glucose:solanidine glucosyltransferase and UDP-galactose:solanidine galactosyltransferase from potato tuber. *Plant Sci.* **84:**34–44.
20. Lüning, H. U. and E. Schlösser. 1975. Role of saponins in antifungal resistance. V. Enzymatic activation of avenacosides, *Z. Pflanzenkrankheit Pflanzenschutz* **82:**699–703.
21. Nisius, H. 1988. The stromacentre in Avena plastids: an aggregation of β-glucosidase responsible for the activation of oat-leaf saponins. *Planta* **173:**474–481.
22. Inoue, K. and Y. Ebizuka. 1995. Purification and characterization of a β-glucosidase which converts furostanol glycosides to spirostanol glycosides from *Costus speciosus.* In *Saponins: Chemistry and Biological Activity,* eds. G. Waller and K. Yamasaki, in press, Plenum, New York.
23. Gus-Mayer, S., H. Brunner, H. A. W. Schneider-Poetsch, F. Lottspeich, C. Eckerskorn, R. Grimm, and W. Rüdiger. 1994. The amino acid sequence previously attributed to a protein kinase or a TCP1–related molecular chaperone and co-purified with phytochrome is a β-glucosidase. *FEBS Lett.* **347:**51–54.
24. Gus-Mayer, S., H. Brunner, H. A. W. Schneider-Poetsch, and W. Rüdiger. 1994. Avenacosidase from oat: purification, sequence analysis and biochemical characterization of a new member of the BGA family of β-glucosidases. *Plant Mol. Biol.* **26:**909–921.
25. Lenman, M., A. Falk, J. Xue, and L. Rask. 1993. Characterization of a *Brassica napus* myrosinase pseudogene: myrosinases are members of the BGA family of β-glycosidases. *Plant Mol. Biol.* **21:**463–474.
26. Oxtoby, E., A. Dunn, A. Pancoro, and M. A. Hughes. 1991. Nucleotide and derived amino acid sequence of the cyanogenic β-glycoside (linamarase) from white clover (*Trifolium repens* L.). *Plant Mol. Biol.* **17:**209–219.
27. Agarwal, S. K. and R. P. Rastogi. 1974. Triterpenoid saponins and their genins. *Phytochemistry* **13:**2623–2645.
28. Chandel, R. S. and R. P. Rastogi. 1980. Triterpenoid saponins and sapogenins: 1973–1978. *Phytochemistry* **19:**1889–1908.
29. Mahato, S. B. and A. K. Nandy. 1991. Triterpenoid saponins discovered between 1987 and 1989. *Phytochemistry* **30:**1357–1390.

30. Mahato, S. B., S. K. Sarkar, and G. Poddar. 1988. Triterpenoid saponins. *Phytochemistry* **27**:3037–3067.

31. Mahato, S. B., A. N. Ganguly and N. P. Sahu. 1982. Steroid saponins. *Phytochemistry* **21**:959–978.

32. Crombie, L., W. M. L. Crombie and D. A. Whiting. 1984. Structures of the four avenacins, oat root resistance factors to "take-all" disease. *J. Chem. Soc., Chem. Communic.* **4**:246–248.

33. Crombie, L., W. M. L. Crombie and D. A. Whiting. 1986. Structures of the oat root resistance factors to take-all disease, avenacins A-1, A-2, B-1 and B-2 and their companion substances. *J. Chem. Soc., Perkins* **I**:1917–1922.

34. Tschesche, R., M. Tauscher, H. W. Fehlhaber and G. Wulff. 1969. Avenacosid A, ein bisdesmosidisches Steroidsaponin aus *Avena sativa. Chem. Berichte* **102**:2072–2082.

35. Tschesche, R. and P. Lauven. 1971. Avenacosid B, ein zweites bisdesmosidisches Steroidsaponin aus *Avena sativa. Chem. Berichte* **104**:3549–3555.

36. Goodwin, R. H. and B. M. Pollock. 1954. Studies on roots. I. Properties and distribution of fluorescent constituents in *Avena* roots. *Am. J. Bot.* **4**:516–520.

37. Maizel, J. V., H. J. Burkhardt and H. K. Mitchell. 1964. Avenacin, an antimicrobial substance isolated from *Avena sativa.* I. Isolation and antimicrobial activity. *Biochemistry* **3**:424–431.

38. Crombie, W. M. L. and L. Crombie. 1986. Distribution of avenacins A-1, A-2, B-1 and B-2 in oat roots: Their fungicidal activity towards "take-all" fungus. *Phytochemistry* **25**:2069–2073.

39. Turner, E. M. 1953. The nature of resistance of oats to the take-all fungus. *J. Exp. Bot.* **4**:264–271.

40. Osbourn, A. E., B. R. Clarke, P. Lunness, P. R. Scott and M. J. Daniels. 1994. An oat species lacking avenacin is susceptible to infection by *Gaeumannomyces graminis* var. *tritici. Physiol. Mol. Plant Pathol.* **45**:457–467.

41. Courtney, W. H. and V. N. Lambeth. 1977. Glycoalkaloid content of mature fruit of *Lycopersicon* species. *HortScience* **12**:550–551.

42. Juvik, J. A. and M. A. Stevens. 1982. Inheritance of foliar α-tomatine content in tomatoes. *J. Am. Soc. Hort. Sci.* **107**:1061–1065.

43. Juvik, J. A., M. A. Stevens and C. M. Rick. 1982. Survey of the genus *Lycopersicon* for variability in α-tomatine content. *HortScience* **17**:764–766.

44. Rick, C. M., J. W. Uhlig and A. D. Jones. 1994. High α-tomatine content in ripe fruit of Andean *Lycopersicon esculentum* var. *cerasiforme:* Developmental and genetic aspects. *Proc. Nat. Acad. Sci.,* USA **91**:12877–12881.

45. Prokoshev, S. M., E. I. Petrochenko and V. A. Paseshnichenko. 1956. Tomatinase in tomato leaves. *Dokl. Akad. Nauk SSSR* **106**: 313.

46. Tsukamoto, C., A. Kikuchi, K. Harada, K. Kitamura, and K. Okubo. 1993. Genetic and chemical polymorphisms of saponins in soybean seed. *Phytochemistry* **34**:1351–1356.

47. Steel, C. S. and R. B. Drysdale. 1988. Electrolyte leakage from plant and fungal tissues and disruption of liposome membranes by α-tomatine. *Phytochemistry* **27**:1024–1030.

48. Arneson, P. A. and R. D. Durbin. 1968. The sensitivity of fungi to α-tomatine. *Phytopathology* **58:**536–537.

49. Schlösser, E. 1972. Sterol dependent membranelytic action of saponins. *Phytopathol. Z.* **74:**91–94.

50. Olsen, R. A. 1973. Triterpeneglycosides as inhibitors of fungal growth and metabolism. 6. The effect of aescin on fungi with reduced sterol contents. *Physiol. Plant.* **29:**144–149.

51. Défago, G. and H. Kern. 1983. Induction of *Fusarium solani* mutants insensitive to tomatine, their pathogenicity and aggressiveness to tomato fruits and pea plants. *Physiol. Plant Pathol.* **22:**29–37.

52. Kesselmeier, J. and B. Urban. 1983. Subcellular localization of saponins in green and etiolated leaves and green protoplasts of oat (*Avena sativa* L.). *Protoplasma* **114:**133–140.

53. Roddick, J. G. 1978. Effect of α-tomatine on the integrity and biochemical activities of isolated plant organelles. *J. Exp. Bot.* **29:**1371–1381.

54. Schulz, G. and H. Sander. 1957. Über cholesterin-Tomatid. Eine neue Molekülverbinding zur analyse und preparativen Gewinnung von Steroiden. *Hoppe Seylers Z. Physiol. Chem.* **308:**122.

55. Roddick, J. G. and R. B. Drysdale. 1984. Destabilization of liposome membranes by the steroidal glycoalkaloid α-tomatine. *Phytochemistry* **23:**543–547.

56. Fewell, A. M. and J. G. Roddick. 1993. Interactive antifungal activity of the glycoalkaloids α-solanine and α-chaconine. *Phytochemistry* **33:**323–328.

57. Takechi, M., S. Shimada and Y. Tanaka. 1991. Structure-activity relationships of the saponins dioscin and dioscinin. *Phytochemistry* **30:**3943–3944.

58. Roddick, J. G., A. L. Rijnenberg, and M. Weissenberg. 1990. Membrane-disrupting properties of the steroidal glycoalkaloids solasonine and solamargine. *Phytochemistry* **29:**1513–1518.

59. Osbourn, A. E. 1996. Saponins and plant defence - A soap story. *Trends Plant Sci.,* **1:**4–9.

60. Turner, E. M. 1961. An enzymic basis for pathogen specificity in *Ophiobolus graminis. J. Exp. Bot.* **12:**169–175.

61. Crombie, W. M. L., L. Crombie, J. B. Green, and J. A. Lucas. 1986. Pathogenicity of take-all fungus to oats: its relationship to the concentration and detoxification of the four avenacins. *Phytochemistry* **25:**2074–2083.

62. Osbourn, A. E., B. R. Clarke, J. M Dow and M. J. Daniels. 1991. Partial characterization of avenacinase from *Gaeumannomyces graminis* var. *avenae. Physiol. Mol. Plant Pathol.* **38:**301–312.

63. Bowyer, P., B. R. Clarke, P. Lunness, M. J. Daniels and A. E. Osbourn. 1995. Host range of a plant pathogenic fungus determined by a saponin detoxifying enzyme. *Science* **267:**371–374.

64. Lüning, H. U., B. G. Waiyaki, and E. Schlösser. 1978. Role of saponins in antifungal resistance VIII. Interactions *Avena sativa-Fusarium avenaceum. Phytopathol. Z.* **92:**338–345.

65. Lüning, H. U. and E. Schlösser. 1976. Role of saponins in antifungal resistance VI. Interactions *Avena sativa-Drechslera avenacea. J. Plant Disease Protection* **83:**317–327.

66. Arneson, P. A. and R. D. Durbin. 1967. Hydrolysis of tomatine by *Septoria lycopersici;* a detoxification mechanism. *Phytopathology* **57:**1358–1360.

67. Pegg, G. F. and S. Woodward. 1986. Synthesis and metabolism of α-tomatine in tomato isolines in relation to resistance to *Verticillium albo-atrum. Physiol. Mol. Plant Pathol.* **28:**187–201.

68. Ford, J. E., D. J. McCance and R. B. Drysdale. 1977. The detoxification of α-tomatine by *Fusarium oxysporum* f.sp. *lycopersici. Phytochemistry* **16:**544–546.

69. Verhoeff, K. and J. I. Liem. 1975. Toxicity of tomatine to *Botrytis cinerea,* in relation to latency. *Phytopathol. Z.* **82:**333–338.

70. Osbourn, A. E., P. Bowyer, P. Lunness, B. Clarke, and M. Daniels. 1995. Fungal pathogens of oat roots and tomato leaves employ closely related enzymes to detoxify host plant saponins. *Mol. Plant-Microbe Interact.* **8:**971–978.

71. Sandrock, R. W., D. DellaPenna and H. D. VanEtten. 1995. Purification and characterization of β_2-tomatinase, an enzyme involved in the degradation of α-tomatine and isolation of the gene encoding β2–tomatinase from *Septoria lycopersici. Mol. Plant-Microbe Interact.* **8:**960–970.

72. Arneson, P. A. and R. D. Durbin. 1968. Studies on the mode of action of tomatine as a fungitoxic agent. *Plant Physiol.* 4**3:**683–686.

73. Asher, M. J. C. and P. J. Shipton, eds. 1981. *Biology and Control of Take-all,* Academic Press, London.

74. Scott, P. R. 1981. Variation in host susceptibility. In *Biology and Control of Take-all,* eds. M. J. C. Asher, and P. J. Shipton, pp. 221–236. Academic Press, London.

75. Turner, E. M. 1940. *Ophiobolus graminis* Sacc. var. *avenae* var. N., as the cause of take all or whiteheads of oats in Wales. *Trans. Br. Mycol. Soc.* **24:**269–281.

76. Osbourn, A. E. Saponin detoxification by plant pathogenic fungi. In *Saponins: Chemistry and Biological Activity,* eds. G. Waller and K. Yamasaki, in press, Plenum, New York.

77. Heath, M. C. 1991. The role of gene-for-gene interactions in the determination of host species specificity. *Phytopathology* **81:**127–130.

78. Henrissat, B. 1991. A classification of glycosyl hydrolases based on amino acid sequence similarities. *Biochem. J.* **280:**309–316.

79. Henrissat, B. and A. Bairoch 1993. New families in the classification of glycosyl hydrolases based on amino acid sequence similarities. *Biochem. J.* **293:**781–788.

80. Bryan, G. T., M. J. Daniels and A. E. Osbourn. 1995. Comparison of fungi within the Gaeumannomyces-Phialophora complex by analysis of ribosomal DNA sequences. *Appl. Environ. Microbiol.* **61:**681–689.

81. Bryan, G. T. 1995. The *Gaeumannomyces-Phialophora* complex - Molecular taxonomy and characterisation of avenacinase-like proteins. Ph.D. dissertation, University of East Anglia, Norwich, U.K.

82. Urbasch, I. 1986. Transformation von α-tomatin durch *Botrytis cinerea. Planta Med.* **2:**114–118.

CHAPTER

5

The Biology and Molecular Epidemiology of the Geminiviridae Subgroup III

J. K. Brown

INTRODUCTION

The dicot-infecting, whitefly-transmitted (WFT) geminiviruses have emerged as important plant viral pathogens during the last decade, resulting in pandemic disease conditions in subtropical and fringe-temperate locales. Substantial quantities of food and fiber are grown for fresh market and/or for processing in these areas, and these commodities are destined for either local consumption or for export. A heavy reliance on agricultural commodities for economic prosperity in many subtropical areas, and on a surplus production and subsequent export from certain subtropical and temperate locales, collectively, emphasize the serious threat posed by unprecedented geminivirus epidemics in a wide range of food and fiber crops on a global basis. Hence, reducing crop damage and economic losses due to WFT geminivirus epidemics has far reaching implications toward the sustainability of agriculture as a vital link toward creating opportunities and improving the quality of life worldwide.

The increasing negative impact of WFT geminiviruses on contemporary agro-ecosystems is due to a series of well orchestrated, but as yet poorly understood interactions between the geminivirus pathogens, the whitefly vector, and plant species that serve as hosts of both geminiviruses and/or the insect vector.[1,2,3,4,5] A challenge now exists to explore and ultimately to understand the genetic and phenotypic variation within geminivirus-whitefly vector-host plant complexes and the underlying basis of the respective interactions. Studies of the underlying interactions between members of the complex at whole organismal, cellular, and molecular levels are also essential to achieving a greater understanding of the factors responsible for the emergence of new viral pathogens. And, it is vital that these questions also be explored in the context of a molecular epidemiological perspective.

By one definition, molecular epidemiology of virus diseases is the comparative identification and the tracking of the predominant viral species, or "quasi-species", within and across geographic regions from the respective repertoire of host plants in time and space.[6,7] Understanding the molecular epidemiology will require a detailed understanding of the ecology and natural history of organisms involved, combined with a sound knowledge of the genetic structures of virus, vector, and host plant populations involved.[8] Synthesis of the knowledge gained from these studies can ultimately be constructed in the context of the evolutionary biology of the complex, from which the molecular epidemiological hypotheses can be formulated and tested.[6,9,10] The ultimate goal is to obtain sound biological information about the dynamics and interactions of the components of this complex and, ultimately, to utilize these data in a predictive capacity to reduce losses due to disease. Crop loss may be reduced by selecting against virus virulence and/or by decreasing the frequency of vector mediated spread of geminiviral pathogens such that fewer plants are infected with a strain of any given virulence. Although reaching this goal likely requires the use of multiple crop management approaches along with the cultivation of virus resistant cultivars adapted to the particular locale, little is really known about the efficacy of traditional approaches to disease control with respect to geminivirus diseases.

This review begins with a brief epidemiological history of WFT geminivirus incited diseases. Relevant background information is provided as a basis for detailed discussions of the biology and/or molecular biology of the individual components of geminivirus whitefly vector host plant complexes. As with other vector parasite complexes, each biological component of the complex is a member of a different kingdom of organisms or is a distinct life form. This section is followed by a discussion of specific interactions between members of this complex that ultimately lead to the development of disease. Here, emphasis is placed on virus vector and vector host interactions that may provide clues toward the development of novel management strategies that rely on interference with key processes or specific biological interactions. The WFT geminiviruses are then explored in the context of the virus quasi-species concept, underscoring the need to identify and track the prominent viral species within agroecosytems, as essential requisites to determining the specific geminiviruses against which control strategies should be targeted. Finally, several disease management approaches are presented.

BACKGROUND AND HISTORY

Whitefly-transmitted (WFT) geminiviruses have been known as serious pathogens in tropical agricultural systems between the 30 degree parallels north and south since the turn of the century. However, the subgroup III geminiviruses have not been recognized as emerging viral pathogens in agroecosystems, worldwide, until quite recently.[1,2,3,4,5,11,12,13,14,15,16,17] Seminal studies conducted in subtropical and temperate regions during the 1970's revealed that WFT geminiviruses had a unique paired particle morphology and a single-stranded DNA (ssDNA) genome, making them members of a newly discovered group of plant viruses.[18,19,20,21,22] The geminivirus nucleoprotein appears by transmission electron microscopy as a pair of adjoined isometric particles (T=1 capsid structure), and each geminate virus particle (Figure 5–1) contains a single molecule of circular, single-stranded DNA

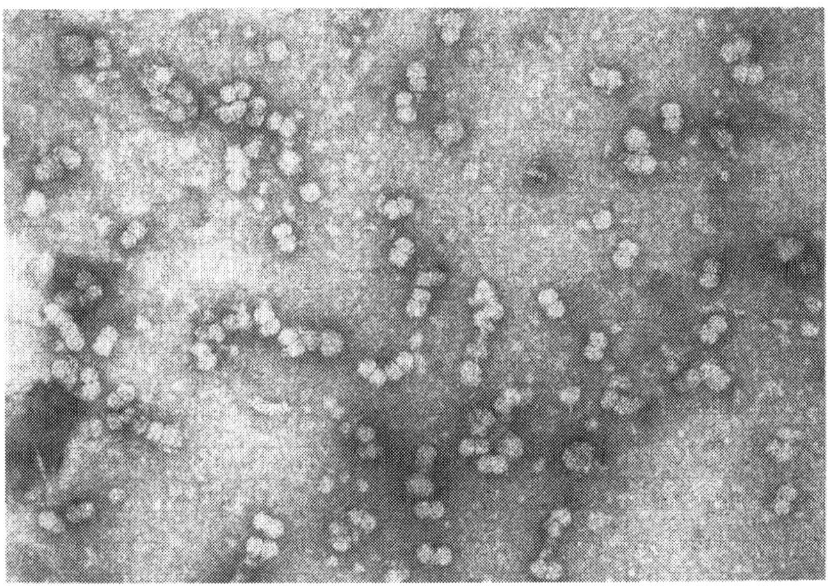

Figure 5–1. Characteristic geminivirus particles (20 X 30 nm) of African cassava mosaic virus (ACMV) by transmission electron microscopy.

Courtesy P. Jones, Rothamsted Experiment Station, Harpenden, Herts, UK.

(ssDNA). Either one or two separately packaged ssDNA molecules are required for infection by the monopartite and bipartite viruses, respectively.[19,20,23,24,25,26] These characteristics, along with their exclusive transmissibility by the whitefly *B. tabaci*, has set the subgroup III geminiviruses apart from other plant viruses described to date. Now known as the subgroup III geminiviruses, these and two subgroups of non-whitefly transmitted geminiviruses were officially recognized as the Geminiviridae by the ICTV in 1978, representing a new family of DNA containing plant viruses.[27] Characterization of WFT geminiviruses has proceeded at a persistent rate since that time, and there is now considerable knowledge about geminivirus distribution, host range phenotypes, virus-vector relationships, genome organization, and gene function for a number of WFT geminiviruses.[2,3,4,12,28,29,30,31,32,33,34]

CLASSIFICATION AND NOMENCLATURE OF THE GEMINIVIRIDAE

Subgroup III geminiviruses are distinguished from other geminiviruses by their requirement for whitefly-mediated transmission and the ability to infect only dicotyledonous plant hosts. A current, but not entirely comprehensive listing of WFT geminiviruses described to date, the respective acronyms, and key reference(s) are provided in Table 5–1. Two other virus subgroups are placed in the

Table 5–1. Subgroup III Geminiviruses and Acronyms

Geminivirus	Acronym	References
Abutilon mosaic virus-West Indies	AbMV-WI	[276,277]
Acalypha yellow mosaic virus	AYMV	Chenulu & Phatak, 1965 in: [1]
Ageratum yellow vein mosaic	AYVV	[17]
African cassava mosaic virus-Nigeria	ACMV-Ni	Storey & Nichols, 1938 in: [1,22]
		[278]
		[22]
African cassava mosaic virus-Kenya	ACMV-Ke	
Asystasia golden mosaic virus	AGMV	[78]
Bean calico mosaic virus	BCMoV	[241]
Bean dwarf mosaic virus-Colombia	BDMV-Co	[279]
Bean golden mosaic virus-Puerto Rico	BGMV-PR	[116,239]
		[238]
Bean golden mosaic virus-Brazil	BGMV-BZ	
Bhendi yellow vein mosaic virus	BYVMV	Uppal et al., 1940 in: [1]
Chino del tomate virus	CdTV	[280]; Gallegos, 1978 in: [2]

Table 5–1. (*Cont.*) Subgroup III Geminiviruses and Acronyms

Geminivirus	Acronym	References
Cotton leaf crumple virus	CLCV	[244,281]
Cotton leaf curl virus	CLCuV	Kirkpatrick, 1930 in: [1] [282]
Cotton mosaic virus	CotMV	Nour, 1959 in: [1]
Cowpea golden mosaic virus	CPGMV	Nene, 1972 in: [1]
Croton yellow vein mosaic virus	CYVMV	[283]
Dolichos yellow mosaic virus	DoYMV	[283]
Eggplant yellow mosaic virus	EYMV	Capoor & Varma, 1950 in: [1]
Eupatorium yellow vein virus	EpYVV	[284]
Euphorbia mosaic virus	EuMV	Costa & Bennett, 1950 in: [1,56]
Florida cabbage virus	FCV	[285,286]
Honeysuckle yellow vein mosaic virus	HYVMV	[287]
Horsegram yellow mosaic virus	HgYMV	[283]
Indian cassava mosaic virus	ICMV	[283]
Jatropha mosaic virus	JMV	[196,249]
Lima bean golden mosaic virus	LGMV	Capoor & Varma, 1948 in: [1]
Malvastrum yellow vein mosaic	MalYVM	Rao & Varma, 1961 in: [1]
Merremia yellow mosaic virus	MerYMV	[205]
X 'Mungbean yellow mosaic virus	MYMV	[1,288]
Okra leaf curl virus	OLCV	[283]
Pepper huasteco virus	PHV	[289]
Pepper mild tigre virus	PepMTV	[290]
Potato yellow mosaic virus	PYMV	Wolf et al., 1949 in: [1,291]
Rhynchosia mosaic virus	RhMV	Bird, 1962 in: [12]
Serrano golden mosaic virus	SGMV	[246]
Sida golden mosaic virus	SiGMV	[292]
Sida yellow vein virus	SYVV	[78]
Sinaloa tomato leaf curl virus	STLCV	[38]
Soybean crinkle leaf virus	SCLV	[293]
Squash leaf curl virus	SqLCV	[43,69,255,257,258,294]
Texas pepper geminivirus	TPGV	[295]
Tobacco leaf curl virus	TLCV	[296]; Storey, 1932 in:[1]
Tomato golden mosaic virus	TGMV	[21,86]
Tomato leaf curl virus-Australia	ToLCV-Au	[84]
Tomato leaf curl virus-India	ToLCV-In	[283]
Tomato leaf curl virus-Sudan	ToLCV-Su	[297]
Tomato mottle virus	ToMoV	[39,298,299]
Tomato yellow leaf curl virus-Israel	TYLCV-Is	[25]
Tomato yellow leaf curl virus-Sardinia	TYLCV-Sr	[85]
Tomato yellow leaf curl virus-Thailand	TYLCV-Th	[95]
Tomato yellow leaf curl virus-Yemen	TYLCV-Ye	[78,300]
Tomato yellow mosaic virus	ToYMV	[61]
Watermelon chlorotic stunt virus	WmCSV	[300]

Geminiviridae: Subgroup I which contains the leafhopper-transmitted, monocotyledonous infecting viruses, and Subgroup II members that are also transmitted by leafhoppers, but infect only dicotyledonous plants. It has been proposed that members of Subgroup III arose by rearrangement and reapportioning of genes from a monopartite, leafhopper-transmitted, geminivirus progenitor genome to either a bipartite or monopartite organization. In contrast, subgroup II contains those with intermediate characteristics that retained determinants of leafhopper transmissibility but have evolved to infect dicotyledonous plants.[33,35] This latter hypothesis is borne out by the observation that Subgroup II members have a gene arrangement more like those of Subgroup III viruses and infect only dicots, but are transmitted by leafhoppers instead of whiteflies.

There are two recently proposed schemes for adding precision to the existing subgroup nomenclature (Table 5–2). One proposed system (Table 5–2A) defines two subfamilies for the Geminiviridae: the Alphageminivirinae (presently Subgroup I) and the Betageminivirinae (presently Subgroups II and III). Within the Alphageminivirinae, is the genus Monogeminivirus (presently Subgroup I) containing the leafhopper-transmitted, monocot infecting virus species (Maize Streak Virus, type member). The Betageminivirinae contains the three genera: the Intergeminivirus, Archeogeminivirus, and Neogeminivirus. Within the Intergeminivirus genus are the leafhopper-transmitted, dicot infecting virus species (beet curly top virus (BCTV), type member). Contained within the Archeogeminivirus genus are the Old World, whitefly-transmitted viruses of dicots with monopartite or bipartite genomes (African cassava mosaic virus, type member), while the Neogeminivirus genus houses the New World, whitefly-transmitted, dicot-infecting species, to date all with bipartite genomes (bean golden mosaic virus, type member).[36]

A second proposed scheme (Table 5–2B) is based on gene by gene phylogenetic analysis of DNA sequences to delineate three genera within the Geminiviridae (as does the existing system), thus, constituting a similar, but somewhat more simplified scheme than that proposed by Padidam et al.[36] This organization scheme groups viruses into (1) the Monogeminiviruses containing the presently recognized subgroup I viruses, or those having monocot hosts and leafhopper vectors, (2) the Bigeminiviruses containing the subgroup III, or those viruses having dicot hosts, a whitefly vector, and either a monopartite or bipartite genome, and (3) the Intergeminiviruses now known as Subgroup II members, which have dicot hosts and a leafhopper vector.[37]

Regardless of the system used, there are few disagreements with respect to phylogenetic relationships and to organization of viruses within these genera. However, establishing additional subtaxa within the Geminiviridae is considered premature in the absence of specific criteria for distinguishing between strains and species.[36] Although valid arguments exist in favor of a standardized system, consideration must be given also to the potential for confusion, relative to published information in which the older terminology is used.

Table 5–2. Two Proposed Systems Of Nomenclature And Phylogenetic Organization Of Geminiviridae

	A. Family: **GEMINIVIRIDAE**[a]			
	Subfamily : Alphageminiviridae (presently subgroup I)		**Subfamily:** Betaviridae (presently subgroups II and III)	
	Genera: *Monogeminiviridae*	*Intergeminiviridae*	Genera: *Archaeoviridae*	*Neogeminiviridae*
	Species	Species	Species	Species
Type:	Maize streak	Beet curly top mosaic	African cassava mosaic	Bean golden
Host:	Monocots	Dicots	Dicots	Dicots
Vector:	Leafhoppers several spp.	Leafhoppers several spp.	Whitefly *Bemisia tabaci*	Whitefly *B. tabaci*
Geography:	Old World	New World	Old World	New World

	B. Family: **GEMINIVIRIDAE**[b]		
Subfamily:	**Monogeminiviridae** (subgroup I)	**Intergeminiviridae** (subgroup II)	**Bigeminiviridae** (subgroup III)
	Species	Species	Species
Type:	Maize streak	Beet curly top	African cassava mosaic
Host:	Monocots	Dicots	Dicots
Vector:	Leafhoppers several spp.	Leafhoppers several spp.	Whitefly *Bemisia tabaci*
Geography:	Old World	New World	Old & New World

[a] 36
[b] 35

GEMINIVIRUS DISEASES

Symptomatology And Host Effects

In general, WFT geminiviruses are virulent pathogens of cultivated crops, with yield losses potentially reaching 80–90%. Although weed hosts develop foliar symptoms, and at times exhibit reduced growth, the effects of geminiviruses on weeds generally appear to be less severe than those observed in cultivated hosts. Foliar symptoms expressed by geminivirus-infected plants are typically either a golden mosaic or bright yellow and green calico, or severe leaf curling and stunting, or combinations thereof. Disease symptoms are also manifest as shortening of internodes (stems and petioles), severe stunting, increased production of flowers, possibly indicating premature senescence, reduced fruit set, and poor size and quality of fruiting structures.

Several symptomless hosts have been identified for geminiviruses in which by definition, the virus is present as is demonstrated by recovery of the virus from inoculated plants. Examples of this phenomenon include eggplant for Sinaloa tomato leaf curl,[38] soda apple for tomato mottle virus[39,40] and the weed *Cynanchum acutum* which is an important reservoir of tomato yellow leaf curl in Israel.[41]

Host Localization and Cytopathic Effects

Whether a particular virus and/or host plant gene(s) facilitates or prevents the invasion of specific cell and tissue types by geminiviruses is not known. For example, there is yet no satisfactory explanation as to why the tobacco species, *Nicotiana benthamiana,* is an almost universally permissive experimental host of geminiviruses when mechanically inoculated with virions, and viral DNA or infectious clones. Why most geminiviruses are capable of experimentally infecting *N. benthamiana*, when the whitefly vector dies following a brief acquisition-access period on the leaves of *N. benthamiana*, is not known. Perhaps this tobacco species has unique genetic characteristics, that, in conjunction with compatible viral genes, permits the (nearly) universal replication and movement of geminiviruses. For this reason, *N. benthamiana* has been the host of choice in studies of genes involved in gene function and replication,[42] and in host range determinants, movement, and pathogenicity.[43,44,45]

Infections by geminiviruses result in alterations of plant cells and organelles, and in the appearance of virus associated structures in infected plants. The most common pathologies are nuclear inclusion bodies, ordered aggregates containing virus-like particles, referred to as strings of beads, and fibrillar rings.[46,47,48,59,50,51] These structures are present in the nuclei of phloem palisade and spongy parenchyma cells and companion cells, and in differentiating and mature sieve elements.[52] Infected nuclei contain nucleoli that are much enlarged and segregated, and appear to be a primary site of pathological changes.[51,52,53,54,55] Viral inclusion bodies are surrounded by endoplasmic reticulum, golgi bodies, and ribosomes and are observed in the cytoplasm of affected cells.[56] Necrosis and degeneration of vascular parenchyma and sieve elements, and cytoplasmic vesiculation is observed,[46,55] indicating extensive damage to the vascular system, expressed by symptoms of severe stunting and reduced growth of infected plants.[57,58,59,60,61,62,63,64] DNA hybridization studies of in Abutilon mosaic virus (AbMV) infected plants demonstrated that viral DNA is only detectable in the nuclei of phloem, and is not present in palisade cells and spongy parenchyma.[65,66]

Interestingly, bean golden mosaic virus (BGMV), which typically produces virus particles and induces cytopathic changes only in phloem-associated cells of bean, has been shown to infect mesophyll and epidermal cells of bean plants coinfected with BGMV and tobacco mosaic virus (TMV) (legume strain).[67] This loss of tissue specificity in the presence of a second unrelated virus suggests that similar interactions may occur in wild type populations in weed and/or cultivated hosts. These observations likely have important evolutionary and epidemiological implications, and collectively, illustrate the importance of understanding the basis for localization of transmissible virus forms in plants, in relation to sites of virus replication and movement, and to those from where the whitefly vector feeds and ingests virus.

Host Range Phenotypes of Subgroup III Geminiviruses

Range is defined here as those plant species that support virus infection, thus, the term is applied to plant species that are naturally occurring hosts, and to those that can be experimentally infected by various means of inoculation.[68] The host ranges of WFT geminiviruses are generally thought to be rather narrow in that they are often confined to one or several plant families. However, as additional viruses are discovered and better characterized, it is clear that geminivirus host range phenotypes span a range from narrow to quite broad. For example, the Caribbean Basin BGMV is highly restricted to legumes as natural hosts. Other examples of geminiviruses with narrow (natural) host ranges are the squash leaf curl virus (SqLCV) from the Western Hemisphere that infects beans and cucurbits,[43,69] and the African cassava mosaic virus (ACMV) from the African continent that infects cassava and *Datura* spp. Among those in a moderate category are the North American cotton leaf crumple virus (CLCV) infecting bean, cotton, *Hibiscus*, and *Malva parviflora L.*,[70] and numerous tomato/pepper infecting geminiviruses of Mexico and the Caribbean Basin (including peripheral areas such as Arizona, Florida, and Texas) that naturally and/or experimentally infect bean, pepper, tomato, and several Malvaceous and Solanaceous weeds.[39] And, several examples of geminiviruses with extremely broad host ranges are the recently described strain of the Middle Eastern tomato yellow leaf curl virus (TYLCV), one of the most virulent WFT geminiviruses in tomato, for which at least six plant families have been identified,[41] and tobacco leaf curl virus from India, for which 28 of 99 species tested were identified as virus hosts, albeit, current thinking is that there may be several distinct viruses involved in this latter etiology.[71]

Host Range Determinants

Complex nature of virus-host interactions presents a challenge to elucidating potentially interactive viral and plant host gene products and other driving forces that govern geminivirus host range phenotype. Studies of host range determinants that influence the host ranges of other plant viruses support the premise that host range phenotypes are the result of compatible interactions involving viral genes and gene products and those of the host plant. Such constraints initially involve host plant interfaces that must be penetrated by virions (cell walls), the ability of the virus to replicate in particular host cell types, and finally, unimpeded cell-to-cell movement and long distance spread.[68] That the vector plays an important role in host selection and in physical placement of WFT geminiviruses by stylet penetration of the epidermis and then phloem cells,[72,73] suggests that genetically based determinants of viral host range phenotypes become relevant only after the virus enters the host cells.

The specific genetic constraints that govern host range of geminiviruses are not well-understood, but virus replication is negligible, or greatly limited in plant species that are considered nonhosts,[74] indicating that replication is a deciding factor in host range determination. The ability to move systemically from sites of replication have also been shown to impose constraints on geminivirus host range.[43,75,76] Although additional studies are needed to understand the basis of geminiviral host range, it can be postulated that gene function(s), and interaction between viral sequences and specific host factors are essential components in determining virus host range phenotypes, reinforcing the involvement of multiple genetic determinants.

Whitefly Vector Host Range Contributions to Disease

Geminiviruses are exclusively pathogens of dicotyledonous plants and typically infect cultivated species and annual and/or suffructescent weeds in temperate and subtropical-tropical locales. Previously, disease outbreaks were sporadic and typically confined to a single or a few host species.[12,77] In retrospect, this may be explicable based on the biotype concept that outbreaks are limited in breadth by host restrictions of the whitefly vector. Ultimately, then, the extent to which virus inoculum is dispersed depends upon host-finding behaviors of the whitefly vector that dictate host preference and host selection.

Host ranges of other plant pathogens are postulated to be limited by, or at least correlated with, the distances that their vectors are capable of dispersing. Perhaps these observations have some bearing on the basis for the relatively narrow host range phenotypes that characterize many WFT geminiviruses. There is increasing evidence of host range differences among populations of the whitefly vector of geminiviruses. Among a suite of *B. tabaci* populations are those with representatively narrow, moderate, and broad host ranges, and distinctive dispersal behaviors.[78,79] Investigations centered around these or other prototype vector populations exhibiting a range of phenotypes may provide insights into the basis for the various host ranges of geminiviruses. Interestingly, the newly discovered B biotype has a broad host range, is capable of transmitting all geminiviruses tested to date, and disperses over long distance.[80] Whether there will be evidence of emerging geminiviruses that have broad host ranges linked to a whitefly vector exhibiting a similar or more broad host range phenotype remains to be seen. Indeed, a strain of TYLCV was described recently that has an extremely broad host range encompassing numerous plant species,[81] compared to the rather narrow host range phenotype of the prototype TYLCV-IS, first described in the 1960s.[82] And, although there is evidence that the B biotype whitefly was present in Israel at least as early as 1985–86,[83] whether a correlation will be found between the recent colonization by a highly polyphagous vector and the emergence of the broad host range strain of TYLCV remains to be demonstrated. Knowledge of the breadth of extremes in *B. tabaci* behavior and in geminivirus host range now provides an opportunity to test this hypothesis with an insect-vectored plant virus.

MOLECULAR BIOLOGY OF THE WFT GEMINIVIRIDAE

Genome Organization and Nomenclature

Subgroup III geminiviruses have either monopartite or bipartite circular, single stranded DNA genomes. The majority of WFT geminiviruses have bipartite genomes, an genome organization found in both Old and New World settings. In contrast, several monopartite viruses have been discovered recently, but this organization is thus far confined to viruses from Old World sites and infect only Solanaceous plant species. Monopartite genomes are approximately 2.7–2.8 kilobases (kb) in size and contain at least six genes encoding proteins of greater than M_r 10,000 Daltons Da.[25,84,85] In contrast, WFT geminiviruses with bipartite genomes are about twice as large, or 5.2 to 5.4 kb, and have two genomic components, designated A and B (also DNA 1 and DNA 2). Each of these components is approximately 2580–2700 bases in size, and collectively, contain six to seven open reading frames (ORFs) of greater than M_r 10,000 Da.[33] Although the monopartite viral genes are arranged on a single component, the number of genes and their respective functions are generally analogous to those found in their bipartite counterparts, and hence the monopartite geminiviruses are also proposed for inclusion as members of Subgroup III.[35,36]

The nomenclature of geminiviral genes and gene products is presently in a state of transition in light of the subsequent emergence of two different working systems. Several proposals made recently regarding gene nomenclature and virus taxonomy are under discussion[35,36] and attempt to take into account newly acquired information and new phylogenetic information as well as the potential discovery of additional open reading frames (ORFs). Illustrated in Figure 5–2 are the two systems currently proposed for viral gene nomenclature of the bipartite members (Fig. 2 A, B), and the system used for the monopartite geminiviruses (Fig. 2 C). For purposes of continuity, the system of gene nomenclature shown in Fig. 2A[86] will be used throughout this chapter.

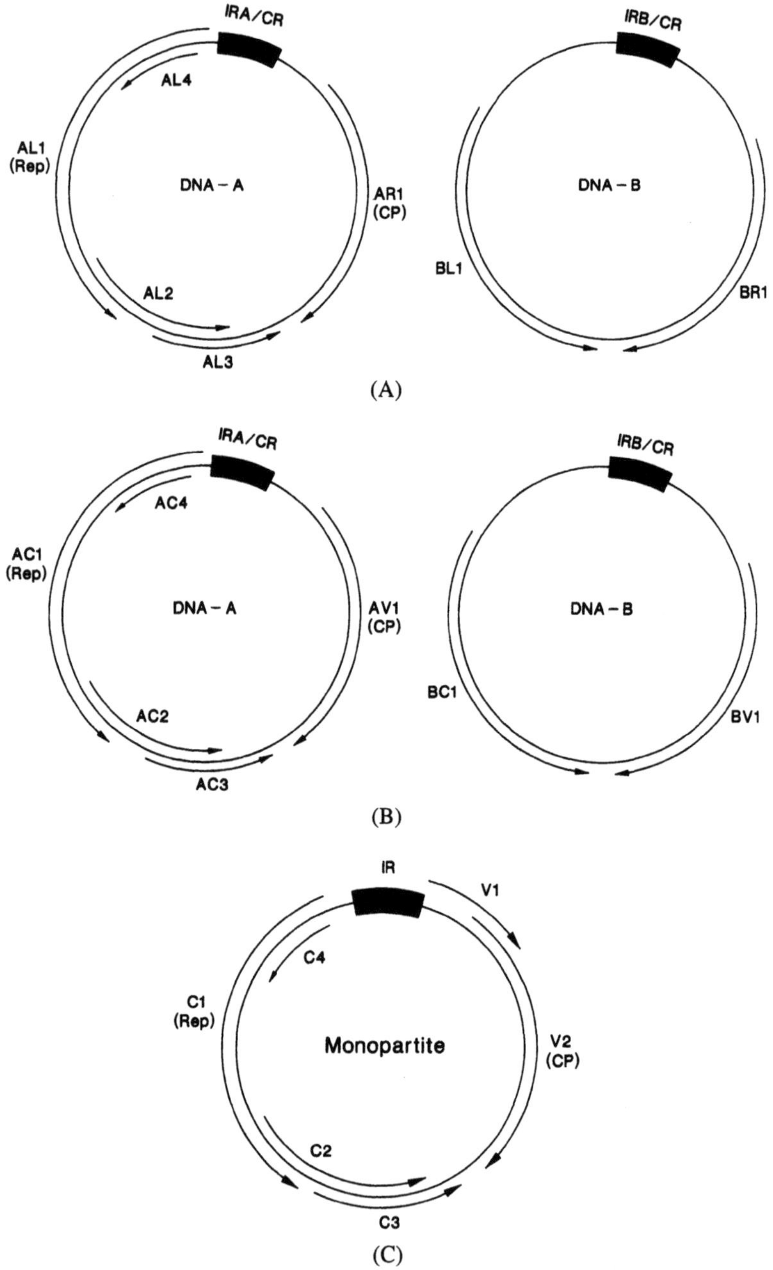

Figure 5–2. Gene arrangement and the two systems of gene nomenclature presently in use for the bipartite geminiviruses (A, B), and gene arrangement and nomenclature of the monopartite geminiviruses (C), Subgroup III of the Geminiviridae.

The Bipartite WFT Geminiviruses

The viral A component encodes all viral functions needed for replication and encapsidation of the ssDNA genome.[87,88,89,90] The B component has two ORFs, BR1 and BL1, both known to provide functions or movement in infected plants.[88,90,91] In bipartite geminiviruses, the A and B components are both required for systemic infection of host plants.[20,26] The A component of the bipartite viruses is considered to have four ORFs based on evidence of functional gene products: one gene transcribed from the positive strand in the rightward or clockwise direction on the circular genome, and termed A-right (R) 1, or AR1 (CP), and four genes transcribed from the negative strand in the leftward or counterclockwise direction, termed A-left (L) or AL1 (Rep), AL2 (TrAp), and AL3 (REn), and AL4.[42,92]

A fifth ORF, AL5, has been predicted for pepper hausteco virus (PHV) and appears to be loosely conserved in several other geminiviruses, but a gene product and specific function have yet to be demonstrated for this putative protein.[93] The bipartite geminiviruses from Old World locations also have a predicted ORF termed the precoat[92], or AR2 (AV1)[94], located in front of the capsid protein gene, but there is no evidence for a viable gene product or of an associated function for this predicted in the bipartite viruses.

The Monopartite WFT Geminiviruses

The Single-component nature and the gene organization of the Subgroup III monopartite geminiviruses were discovered only recently, hence aspects concerning gene organization and function in these viruses have been studied only recently and new information is forthcoming. The gene organization of the monopartite viruses has been shown to be somewhat similar to that found in the bipartite viruses except that all ORFs are arranged within a single component of slightly larger size. These viruses appear to have several genes analogous in size and function to those in the bipartite viruses and several unique ORFs for which function(s) are not yet completely understood. The monopartite geminiviruses have at least six predicted ORFs encoding polypeptides of greater than M_r 9.5 to 10 kDa[25,84,85] that encode products termed V1 and V2 (CP), transcribed in the rightward, clockwise direction (V = viral stand), whereas C1 (Rep), C2, C3, and C4 are transcribed in the leftward or counterclockwise direction (C = complementary strand). V2 (CP), C1 (Rep), C2, and C3 appear to be functionally analogous to AR1(CP), AL1(Rep), AL2, and AL3, respectively,[28] in the bipartite counterparts. In these viruses, a unique ORF, the precoat, designated V1 because it is the first rightward ORF, making the coat protein gene the second, and hence the designation V2 instead of AR1 (AV1) as in the bipartite members (Fig 5– 2A-C). The C4 and V1 gene products of the monopartite viruses appear to be unique and may not have analogous, functional homologues in the bipartite members; instead, these two gene products are thought to be involved in movement, in contrast to the functional analogs, BL1 and BR1 in bipartite members.

In the monopartite group, the capsid protein is also thought essential for viral movement, whereas the capsid protein of the bipartite viruses enhances viral movement but is not absolutely required for systemic spread of the virus. One prospective intermediate virus, the tomato yellow leaf curl virus from Thailand (TYLCV-Th) has two genomic components with an A component having a gene arrangement like the monopartite members and a B component arranged like those of bipartite members. The monopartite-like A genome alone can incite a viable systemic infection of tomato in the absence of the B component. When the A and B components are coinoculated to tomato, symptoms develop slightly more rapidly, and initially appear somewhat more severe, however, within several days, symptoms in infected tomato are indistinguishable from those caused by A alone.[95] Whether the B component enhances viral movement but is not absolutely required for systemic infection by TYLCV-Th because the virus has functional V1 and C4 genes on the A component as in other monopartite viruses has not been addressed experimentally.

The Intergenic Region

A large intergenic region (IR) of approximately 280 to 350 bases is located between the AR1 and AL1 genes on the A-component, and between the BR1 and BL1 genes on the B component, and at the analogous location between C1 and V1 on the monopartite viruses (Figure 5–2 C). Within the IR of bipartite viruses is a region referred to as the common region (CR), that is identical or highly similar for the A and B components but differs to varying degrees between distinct viruses. The IR/CR contains the origin of replication and promoters responsible for the transcription of leftward and rightward gene sequences that bind nuclear host factors.[33,96,97,98,99,100] Located within the IR/CR is an inverted repeat that is capable of forming a hairpin structure, and the loop of the hairpin contains an invariant sequence,TAATATTAC, that is part of the origin of replication.[28,33]

Gene Products and Gene Function

The Capsid Protein (CP)

The AR1 gene encodes the capsid protein gene, or CP (27 to 28 kDa), the most highly conserved gene among the Geminiviridae. The degree of conservation between members of Subgroup III is estimated to range from 67 to 90% similarity at the nucleotide level,[101] and as high as 93 to 95% identity at the amino acid level for some pairwise comparisons. This gene product functions to encapsidate the ssDNA genome, one circular molecule per capsid. It has been known for some time that absence of the AR1 gene product results in a delay or attenuation of disease symptoms,[102] and a concomitant reduction in the amount of ssDNA accumulated in infected plants.[103] In addition, the expression of AR1 is transactivated by the AL2 gene product (transcription factor), as is the expression of BR1.[104] Recent studies suggest that the capsid protein may play an accessory, albeit nonessential role in virus spread within the plant,[105] suggesting multiple functions for AR1 (CP) involving complex interactions with several other viral gene products. In contrast, the capsid protein (V2) is necessary for infectivity of subgroup III monopartite viruses.[106]

The capsid protein has been implicated as a determinant of vector-mediated transmission for both the whitefly and leafhopper vectored members of the Geminiviridae.[31,101,107,108] This suggestion may in part be explained by the high degree of homology[109] observed at the amino acid sequence level among the whitefly-transmitted subgroup whose members are vectored exclusively by the same whitefly species. In contrast, the leafhopper transmitted geminiviruses (subgroups I, II) exhibit lower relative amino acid sequence identities, and are transmitted by several different leafhopper vectors (Figure 5-3).

The presence of predominant, conserved epitopes on the capsid proteins of WFT geminiviruses was also predicted, prior to the availability of nucleotide or deduced amino acids sequence information. This prediction was based on the observation that polyclonal antisera made against purified geminivirus capsid of one virus cross-reacted with all other WFT geminiviruses tested. These results also indicated no antigenic differences were detectable between viruses using polyclonal antibodies and precluded the conventional use of antibodies for virus or strain differentiation.[31] In studies with monoclonal antibodies (Mabs) raised to purified geminivirus capsid protein to an Old World virus, certain Mabs showed a high degree of cross-reactivity to candidate viruses from both Old and New World sources. Among this panel, a single Mab, or several different Mab's used collectively as a subpanel, were identified that permitted differentiation of distinct candidate geminiviruses through epitope profile analysis, but no Mab was found to be virus specific.[110,111,112,113,114]

Predictions about the high degree of amino acid sequence similarity among subgroup III viruses and the subsequent low abundance of (predicted) unique epitopes against which virus-specific antibodies can be raised can now be verified by examination of geminivirus coat amino acid sequences, as predicted from their nucleotide sequences. Indeed, sequence comparisons among a suite of geminiviruses indicate that the N terminus of the molecule is somewhat variable, and in theory, could serve as a source of unique epitopes if oriented optimally for immunological exposure. However, the core, or central region of the molecule, and the predicted C-terminus sequences indicate a high degree of similarity among all subgroup III viruses compared thus far.

In addition, there are several highly conserved regions within the central core of the capsid protein gene, beginning at approximately nt 470 and ending at approximately nt 1070,[115] [using the nt coordinates of the A-component of BGMV-Puerto Rico (BGMV-PR) as the prototype virus sequence, as published in GenBank].[116] Within the core region can be translated three conserved motifs of greater than four amino acids and three motifs of 8 to 10 identical amino acids,

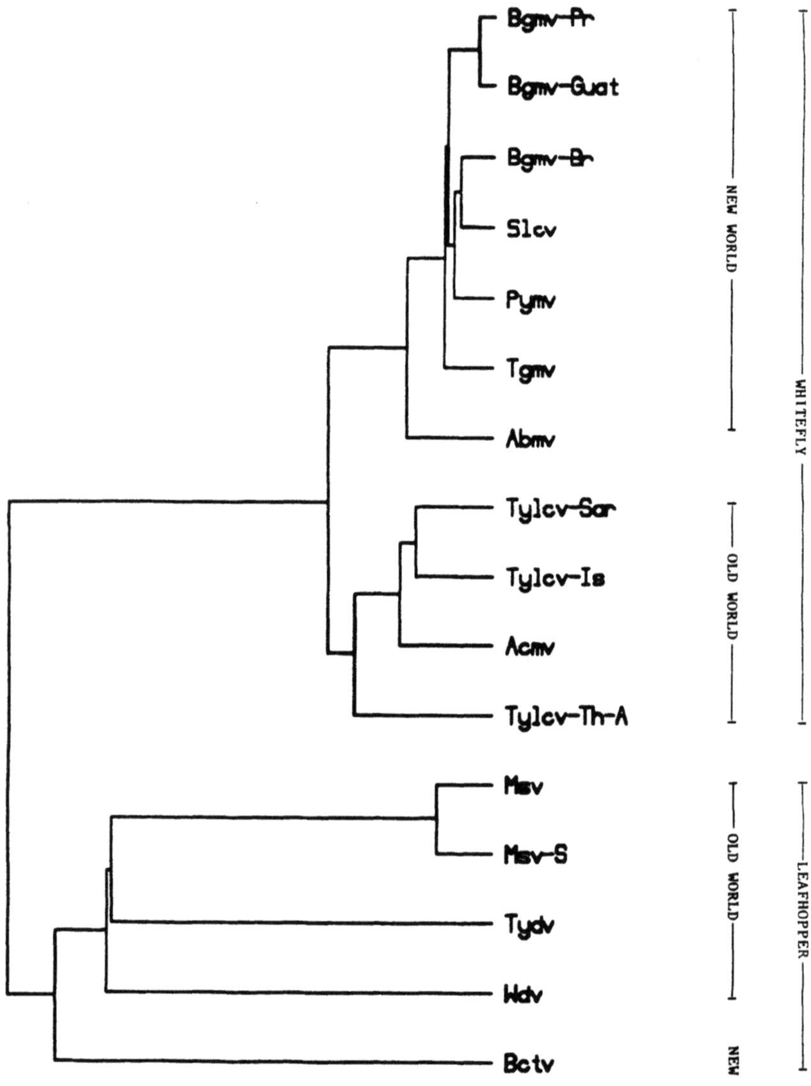

Figure 5–3. Dendrogram illustrating the different subgroups of the Geminiviridae, based on relationships predicted from aligned capsid protein (CP) gene sequences of several well-characterized geminiviruses. Virus clusters correspond to contemporary taxonomy schemes that separate the Geminiviridae into three subgroups based on insect vector specificity, monocot or dicot host range phenotype, and recent geographical affiliation. *Courtesy of Rosell and Brown,*[165]

making the core the most conserved region of the coat protein gene. A second extremely highly conserved motif is quite extensive and begins approximately around the amino acid 212, extending to the end of the C-terminus of the polypeptide. The functional importance of these variable and conserved regions of the coat protein are under investigation in several laboratories.

Replication Protein (Rep)

The AL1 or replication protein (Rep) is the second most highly conserved gene product of the WFT geminiviruses at the nucleotide sequence level (60–71% similarity). This multi-functional protein is the only viral protein absolutely required for replication[87,117,118] which occurs by a rolling circle mechanism.[119,120] Rep proteins appear to be virus-specific at the level of species and/or strain, based on evidence that Rep protein will only support replication of the homologous or a closely related (strain?) virus.[33,121,122] AL1 protein binds to a specific sequence element 50–75 bp upstream of the hairpin[98,100,123] and binding of AL1 to this sequence is required for replication.[123] No polymerase activity has been attributed to this viral protein. Instead, Rep protein introduces a sequence-specific nick in the plus strand to initiate and terminate rolling circle replication.[124] The process of viral DNA replication is presumably carried out by host enzymes. Interestingly, the AL1 binding site is not required for nicking, which occurs between the last AC of the nonanucleotide sequence in the stem loop.[124,125,126] Rep protein also has an ATPase activity required for replication, although the precise role of its activity is not yet clear.[127] Finally, the AL1 gene product acts as a repressor of its own transcription.[128]

Viral regulatory Genes

The AL2, AL3, and AL4 gene products have regulatory functions that facilitate systemic infection in plant hosts. The AL2 or transactivating protein (TrAp)[28] is necessary for efficient expression of the rightward genes, AR1 and BR,[104,129] and is also required for normal accumulation of viral ssDNA.[103,130] This may due to a need for coat protein to stabilize ssDNA by encapsidation. AL2 transactivates the transcription of the rightward genes, AR1 (CP) and BR1.[104] In so doing, the AL2 acts through a sequence element downstream of the hairpin in the common region,[131] although it does not appear to bind the sequence directly. In vascular tissue, AL2 protein may also interact with a second sequence element to derepress expression of the AR1 (CP) gene Presumably, AL2 operates through interactions with sequence-specific host factors that bind viral DNA. Biochemical studies indicate AL2 is a phosphoprotein that binds DNA in a non-specific manner and also binds zinc, suggesting the presence of a zinc finger motif.[131] AL2 function appears not to be viral specific in that heterologous geminivirus AL2 proteins complement a TGMV AL2 mutant.[131]

The AL3 gene product, or replication enhancer (REn) protein,[28] although not essential for replication, has been shown to play an important role in this process.

Genetic evidence indicates that plants infected with AL3 mutants show attenuated symptoms and accumulate up to 50–fold less viral DNA than wild type virus,[92,103,132] leading to the suggestion that this protein functions as a replication enhancement factor.[28] A high degree of conservation of amino acid sequences and functional homology among AL3 proteins has been noted through sequence comparisons of bipartite geminiviruses[133] and through genetic evidence that AL3 function can be complemented by viruses in both Subgroups II and III.[131,134] Precisely how AL3 protein enhances replication remains to be determined.

The function of the AL4 protein has not yet been elucidated. Previous mutational analysis of AL4 ORFs of TGMV and ACMV indicated that the gene product was not required for infectivity and symptom expression.[87,92,135,136] However, recent studies with the analogous gene product C4 of Subgroup III monopartite demonstrated that plants inoculated with C4 mutants produced attenuated symptoms.[137] These results were interpreted to mean that C4 is involved in either virus movement or functions in symptom expression, respectively. There is also evidence that AL4 may be involved in repression of AL1 gene expression.[136]

Movement Functions

That the gene products on the B components of bipartite geminiviruses are needed for systemic infection has been known for some time through studies of laboratory mutants, and the discovery of one movement-deficient isolate from naturally-infected plants.[32,138,139,140,141] Functional and genetic assays with the BR1 and BL1 have shown that these proteins are essential in effecting systemic movement,[91,142,143] for producing disease symptoms in plants,[144,145] and is a determinant of host range of bipartite geminiviruses.[105,138,146,147] Quite recently, the BR1 and BL1 gene products were also shown to have DNA binding activities.[105] Although evidence is yet forthcoming, two recent studies have proposed that BR1 and BL1 act coordinately in viral movement.[105,143] In one study, BR1 appears to be a single-stranded DNA binding protein involved in shuttling the viral ssDNA from the nucleus to the cytoplasm following viral replication.[105]

Tentatively proposed is a model in which the ssDNA-BR1 complex leaves the nucleus for the cytoplasm where it is directed to the plasma membrane through the intervention of BL1. The model infers both cell-to-cell and long distance movement of an unencapsidated nucleoprotein (ssDNA-BR1) complex, whereby the ssDNA is transported to uninfected cells (cell-to-cell) and by way of the phloem (long distance) to initiate infection of new cells and tissues, respectively.[105] BR1 was localized to cell nuclei and BL1 to cell wall and plasma membrane fractions in this study. Using a complementary approach of microinjection into mesophyll cells, Noueiry et al., (1994) suggest that BL1 of bean dwarf mosaic virus (BDMV) can increase the size exclusion limit of plasmodesmata, and have proposed that viral double-stranded DNA (dsDNA) as well as BL1 moves cell-to-cell, although these are issues that remain to be resolved. The BL1 is important in viral pathogenicity for SLCV, and there is recent evidence that the capsid protein also plays a role in systemic movement of the virus, albeit, these interactions are not yet completely understood.[105,147]

In contrast, much less is known about the movement genes of the monopartite WFT geminiviruses. It is clear that V1 and V2 (CP) are essential for systemic spread and evidence is accumulating that C4 may assist in virus movement.[106,137]

Evolutionary divergence has been demonstrated between the movement proteins of Old and New World bipartite viruses. In one study, the B components of putative progenitor Old World (ACMV and ICMV) viruses were effective in potentiating the movement of both Old and New World (TGMV and AbMV) viruses, whereas movement functions of the New World viruses were interchangeable only between other New World viruses.[139] These observations support the prediction based on a distance analysis of B component gene sequences in which New World virus genes appear to be evolutionarily derived from, but are now composed of, a distinct subset of the Old World genes.[35] In contrast, pseudo-recombinant experiments in which the B components of two other New World viruses, TGMV and SqLCV, were interchanged and coinoculated with the heterologous A component, indicated that viable pseudorecombinants were not formed between these two New World virus combinations.[33,139] These latter results suggest additional radiation in the movement proteins of some New World viruses.

The Intergenic Region

The intergenic region of geminiviruses is important in many aspects of the virus life cycle and has been studied in detail only recently. More detailed aspects of viral gene expression, gene regulation, and specific DNA sequences required for replication the subgroup III of the geminiviridae can be found in definitive reviews about these specific subjects.[28,32,33,34,148,149,150] The ssDNA genomes of geminiviruses are replicated and expressed by way of dsDNA intermediates or replicative form (RF) DNA. The first step is the conversion of ssDNA to dsDNA by a priming mechanism. Because ssDNA (plus strand) by itself is infectious, priming and minus strand synthesis must be carried out entirely by host enzymes. This process is not well understood, but there is some evidence that the RNA primer initiates within the intergenic region in the genomes of WFT geminiviruses.[151] In any event, once it is synthesized the dsDNA RF serves as template both for replication and transcription. Further replication, as stated previously, occurs by a rolling circle mechanism and requires for initiation the viral AL1 protein as well as a plus strand origin of replication.

Replication, Transcription, and Translation

The plus strand origin of replication (ori) encompasses roughly 100–bp of AL1–proximal intergenic region sequence[146] and is composed of several distinct elements. One of these is the AL1 binding site.[98,100] It serves as an origin recognition sequence, and binding of AL1 protein to this site is required for replication.[123] In TGMV, the AL1 binding site lies approx 60 bp upstream of the conserved hairpin and consists of two 5 bp direct repeats separated by a 3 bp central core (GGTAGTAAGGTAG).[123] Related but not identical sequences are found at similar locations in other geminiviruses,[96] and as expected AL1 proteins apparently bind only to their homologous and not to heterologous origins.[123] A second required element is the conserved hairpin that contains the invariant sequence TAATATTAC.[152,153] Plus strand synthesis initiates within the hairpin by the introduction of a nick (introduced by AL1 protein) between the T and AC in the conserved nonanucleotide, and the 3' hydroxyl group generated by the nick serves as primer for the synthesis of nascent plus strand DNA.[124,126] The hairpin sequence does not contribute to virus-specific recognition of the origin. Once an origin is recognized and specific nicking occurs, AL1 protein binds to the 5' phosphoryl end at the nick, and synthesis of new plus strand DNA from the 3'-OH terminus is carried out by host DNA polymerase using the circular minus strand as template. The newly synthesized plus strand displaces the previously existing plus strand as synthesis occurs around the circular template. Once the origin of replication is restored to double-stranded form, AL1 protein nicks again to release the unit-length, displaced plus strand, which is simultaneously circularized. The process of initiation and termination are therefore similar, and depend on the topoisomerase-like nicking-closing activity of AL1 protein.[124] The products of one round of replication are a nicked circular dsDNA and a circular ssDNA plus strand. The plus strand can either be primed and re-enter the replication pool, or it can be removed by encapsidation or perhaps by interactions with a movement protein.

The transcriptionally active form of the virus is thought to be the covalently closed circular dsDNA replicative form observed in chromatin-like structures.[154] Transcription is highly regulated and proceeds bidirectionally from the IR/CR at specific transcriptional start sites on the plus (virus sense) and minus (complementary sense) strands of the A and B components for bipartite members, and from plus and minus strands of the single genomic component in the monopartite viruses. Ribonucleic acid polyadenylated transcripts have been identified and mapped for the major ORFs present on the bipartite genomes of tomato golden mosaic geminivirus (TGMV),[155,156,157,158] ACMV,[159] and abutilon mosaic virus (AbMV),[160] and four polycistronic transcripts (accounting for all ORFs) have been identified and mapped on the monopartite genome of tomato leaf curl virus-Australia.[161] Viral transcription is carried out by the host RNA polymerase II, and in the case of most geminivirus promoters, RNA polymerase and host transcription factors are sufficient for expression. However, certain promoters, for example AR1 and BR1, require a viral factor (AL2 protein) for full activity.[104,129] AL1 promoter also is capable of repressing its own expression, presumably at the level of transcription.[128]

In vitro and/or in vivo translation of proteins from several geminiviruses has been demonstrated for five gene products encoded on the A component (AL1, AL2, AL3, AL4, and AR1),[100,118,162] and more recently for two gene products from the B component, BR1 and BL1.[105,143,145] Of interest here is the observation that the AL2 and AL3 proteins are both translatable from a single, bicistronic transcript that begins just upstream of the AL2 ORF and proceeds through it as well as ORF AL3.[100]

EXPERIMENTAL PHYLOGENIES

Genetic Markers and Inferred Phylogeny

The evolution of WFT geminiviruses, as with other viruses and organisms can be inferred by comparative analysis of nucleic acid or amino acid sequence identities, or homology.[163] Alignment of the complete nucleotide sequences of these viruses results in the definitive separation of viral genomes into the three major subgroups (I, II, and II), presently accepted as distinct taxa within the family. The subgroup III members further partition according to putative recent geographic origin, or into Old or New World WFT geminiviruses.[35,36,93,164,165] This kind of comparison is only possible, however, for viruses that have been cloned and the entire DNA sequence obtained, and limits the extent of our knowledge about the many other geminiviruses that exist, but for which DNA sequence information is not yet available.

Molecular Markers For Species Identification

In a recent study, an effort was made to identify a geminiviral DNA sequence, or marker, that would be suitable for making predictions about virus identity and virus relationships in the absence of the complete nucleotide sequence.[36] The DNA sequences of all ORFs and the intergenic regions of well-characterized geminiviruses, for which complete genomic sequences are available, were individually compared. Using DNA sequence alignment and several statistical algorithms, trees were constructed to depict virus relationships[36] that are most logical in the context of biological and genetic perspectives, and thus, by definition, are the most informative. The most informative tree was constructed from a sequence data set beginning within the common region and proceeding through nucleotide 198 of the capsid protein gene (encoding the 66 amino acids at the N-terminus of the polypeptide). The corresponding trees constructed from this alignment permitted the same conclusion about geminivirus relationships at the species level, as does an alignment of full-length viral genomic sequences (2.8 to 5.4 kb).[36] This hypothesis now requires rigorous testing among additional virus isolates, and if results prove consistent for a larger sample size, initial steps toward geminivirus identification at the species level will be greatly simplified.

Capsid Protein Gene Sequences

Relationship dendrograms have been generated based on distance measurements between DNA sequences of the most highly conserved gene, the capsid protein gene. In the hands of several workers, this approach has consistently yielded a separation of subgroup III geminiviruses into two clusters, with partitioning apparently correlated to recent geographic origin, one consisting of viruses from the Old World, and the other with viruses from New World sites, irrespective of the monopartite or bipartite nature of the genome.[35,36,133,166,167] When subgroup I and II viruses are included as outgroups, the geographic separation is maintained, but further separation is noted based on either a leafhopper vector or whitefly vector. These collective relationships are shown for several well-characterized geminiviruses (Figure 5–3).[165]

Rep Protein Sequences

Comparison of Rep protein gene (AL1 or AC1) sequences indicates more complex relationships in that subclustering of viruses occasionally obscures some of the vector-related and/or geographically distinct relationships, as are defined by the capsid protein gene dendrogram. One notable example (at the nucleotide sequence level) is the New World virus pepper hausteco virus (PHV) that falls between a cluster of Old World subgroup III viruses and the leafhopper-transmitted subgroup II beet curly top virus.[37,168] Further, when protein (amino acid level) sequences are compared, PHV falls within the Old World cluster that includes ACMV, TYLCV-Sardinia and TYLCV-Israel, and the New World SqLCV partitions into a separate branch.[167]

Movement Gene Sequences

Phylogenetic analysis of the *BR1* and *BL1* genes of eleven viruses or strains delineate several geographically based clusters of viruses, with a few exceptions.[35,36,168] One analysis predicts that the B component genes of the Old World viruses are effectively ancestral to those of the New World viruses and that the New World viruses form a distinct subgroup within the greater diversity of the Old World viruses.[35] The relationships depicted from an alignment of *BR1* genes indicate a logical Old World/ New World separation for most viruses, with PHV again falling in an intermediate position that bridges the clusters containing two distinct biogeographical genotypes. Trees generated by comparison of *BL1* genes yield a similar arrangement of viruses or isolates, with two lineages diverging within the New World branch, but remaining separated from two Old World viruses. One notable exception here is SqLCV, which is in an intermediate, or bridge position.[35,36,167,168] Similar analysis of the available complete nucleotide sequences of the B component of subgroup III viruses unequivocally separates these viruses into New and Old World groups.[35,36]

Iterated Sequences and the Origin of Replication

A characteristic subset of 8 to 12 repeated nucleotides has been identified in the common region by careful examination of aligned geminiviral DNA sequences. These repeated, or iterated, sequences (repeated within a viral genome and found in other viral genomes) are located in the putative binding site of the AL1 gene product (Rep protein). Biochemical and genetic studies support these same locations as sites essential to virus replication. From these analyses, several putative "families" of iterons have been identified among the 30 dicot-infecting geminivirus sequences examined. Within an iteron family, the number, orientation, and spacing of the repeated sequences is conserved within each of (to date) three lineages of dicot infecting geminiviruses:[96,169] (1) the Old World viruses, (2) all the New World viruses studied to date, excluding the SqLCV complex, but also containing the leafhopper transmitted BCTV, and (3) the SqLCV complex.

Numerous lines of experimental evidence support the involvement of these iterons as a replicon specificity determinant of viral replication. Thus, these iterons, by definition, should be similar for viruses from the same or related lineages and identical within the same viral quasi-species (virus-specific). Identification and comparison of these iterated sequences between viruses, in conjunction with genetically based corroboration of predictions made from cross-replication assays, constitute a useful tool for defining relationships between poorly or uncharacterized virus isolates, based on the ability or inability to transactivate replication.

These conserved, repeated sequences and their role in restriction or facilitation of viral replication between closely related or identical viruses also have important implications for the potential evolution of geminiviruses when they occur in mixed infections. It is not inconceivable that such interactions have been, and will be, important in the evolution of new viruses or strains, thereof.

EMERGING WORKING CONCEPTS OF GEMINIVIRUS SPECIES AND STRAINS

The rapid increase in the number of emerging geminiviruses in many world locales has created the need for rapid virus detection and virus identification. Identification implies the ability to detect and differentiate between viral quasi-species and strains thereof. Because the majority of WFT geminiviruses remain uncharacterized to date despite the application of monoclonal antibody/ELISA, polymerase chain reaction (PCR), and DNA hybridization detection, it is not yet possible to accomplish more than positive detection for most, and thus, still lacking is the ability to positively identify many geminiviruses.[1,110,112,113,170,171,172]

Although the availability of nucleotide sequence data for a growing number of discrete viruses and virus isolates has made possible comparisons of nucleotide and deduced amino acid sequences of viral genes WFT from which relative virus relationships may be examined, yet to be developed is a foundation for a biological and genetic basis for differentiation of virus strains versus virus species. At this time, the International Committee for Virus Nomenclature has accepted no proposal toward a definition for species or strains of viruses within the Geminiviridae; thus, these decisions await much additional predictive and experimental evidence towards a workable strategy.

Transactivation of Replication as a Criterion for Viral Strains

Toward this goal, several approaches are being explored in defining DNA sequences of definitive target regions or "molecular markers" of the geminivirus genome that are informative at the species and subspecies levels. Recent studies in which this objective has been addressed have resulted in the proposal that strains of a distinct quasi-species could be defined based upon the establishment of a viable infection through transactivation of replication between the isolates in question. Or, conversely, the inability of the respective isolates to form pseudorecombinants would be indicative of distinct quasi-species. Thus, the capability of viral AL-1 gene products of the isolates in question to cross-replicate the respective B component would be considered robust evidence for a close evolutionary relationship, and one that in effect could assist in defining a functional quasi-species.

Core Region of the Coat Protein as a Molecular Marker?

An approach toward identifying genetic markers useful for rapid virus identification, and which might also serve as informative sequences for species versus subspecies level discrimination is under investigation. In this study, a comparative alignment of the "core" region of the coat protein gene (defined here as nt 494 to 1048) was accomplished with over 120 geminiviruses/isolates from representative world regions and different host plant species. Distance analysis yielded calculated mean distances ranging from 0.05 to 0.57, a range potentially suggestive of quasi-species to subspecies (strains) level differences. A tree constructed from these data shows that isolates clustered first, based on (purported) Old or New World origin, as also observed when the entire coat protein gene is examined. Next, within the two geographic subgroups, many isolates clustered according to either the host species from which they were initially isolated, and/or from a host species within the same plant family.[115,173] Although it is conceivable that related viruses and strains may share host range phenotypes in common with other closely related viruses and strains, it does not necessarily follow that unrelated viruses will either not share common host range phenotypes, or that closely related viruses are obliged to share common host ranges. Whether similar observations will be substantiated for the majority of isolates examined in the future remains to be seen.

Future Perspectives

Ideal for making strain versus species decisions would be collectively: (1) identification of informative sequences useful as definitive phylogenetic markers, (2) empirical evidence for the complementation of AL-1 through transactivation experiments, and (3) host range and symptom phenotypes, or "phenotypic markers" by which to recognize identical, related, or distinct genotypes. In this light, it must be taken into account that the acceptance of these criteria for discriminating between species and strains of the subgroup III viruses, will require full-length infectious clones of each geminivirus isolate in question. Similar decisions about monopartite viruses will require analogous and creative means of assessment.

THE WHITEFLY VECTOR

Taxonomy

The whitefly, *Bemisia tabaci* (Genn.) is a member of the Aleyrodidae (Order: Hemiptera; Suborder: Homoptera). *B. tabaci* was first described from tobacco in Greece in 1889 and designated *Aleyrodes tabaci*, or the tobacco whitefly.[174] Subsequent identifications of *B. tabaci* in other crops such as cotton led to the assignment of additional vernacular names, such as the cotton whitefly, thus, several common names can be found in the literature reference to *B. tabaci*. Among recognized whitefly species, few specialize on annual herbaceous hosts as does *B. tabaci*. Following the discovery in Florida in 1897 of a whitefly on sweet potato that was considered distinct from *B. tabaci*, this insect was designated *B. inconspicua* and given the common name of sweetpotato whitefly.[175] In time, the difficulties in accurately identifying closely related whiteflies were recognized, a confounding factor being the particular morphological characters used in classical whitefly taxonomy, which are known to vary for the same insect species colonizing different host plants.[176,177,178,179] Subsequently, in 1957, a detailed reexamination of worldwide specimens was undertaken and 19 distinct whitefly species were synonymized into a single taxon, *B. tabaci*.[179]

To address the current confusion resulting from recent evidence of greater biological and genetic variability within the taxon than previously suspected, a *B. tabaci* species complex has been proposed to encompass *B. tabaci* from both New and Old World sites. This proposal also provides an avenue by which the designation of additional species epithets can be avoided until concrete evidence is provided.[180,181,182,183] A second viewpoint is that an additional species should be created to account for at least one population, otherwise recognized as *B. tabaci*,[184,185] hence the designation of the B biotype as *B. argentifolli*[184] with the common name of silverleaf whitefly.[185] Additional studies are needed to define the criteria sorting whiteflies into a *B. tabaci species* complex, and to assist in defining separate species.

Reproduction

B. tabaci typically produces 12 or more generations per year (multivoltine) and females can deposit as many as several hundred eggs in a lifetime under optimal conditions.[186,187] Whiteflies exhibit arrhenotokus parthenogenesis, and thus, produce offspring either with or without fertilization. The ratio of males to females fluctuates during a season. In the absence of fertilization, eggs yield males (1N), whereas mating and fertilization of eggs result in female (2N) offspring.[188] The life cycle involves six stages: egg, three nymphal instars, a fourth nymph or a red-eye nymph stage that ceases feeding and is called a *pupa* (but does not undergo a true metamorphosis), and the adult. Rate of development from egg to adult is positively correlated with temperature, with most rapid development occurring optimally around 28°C, with upper and lower thresholds at 11°C and 33°C, respectively.[189] Also affecting rate of development and fecundity of *B. tabaci* is the particular host plant, and these parameters have recently been shown to differ between biotypes.[78,190,191]

Geminivirus Transmission

B. tabaci is the only recognized whitefly vector of geminiviruses worldwide,[1,2,3,12,17,41,192] indicating a high degree of conservation among whitefly vector genotypes with respect to requirements for a successful transmission event. The number of recognized geminiviruses transmitted by *B. tabaci* continues to rise as this whitefly has increased in abundance in agricultural systems, however, it is not possible to estimate the precise number of uncharacterized geminiviruses in the world. Collectively, *B. tabaci* populations worldwide have a broad host range that includes several hundred host species,[193] making it an ideal vector of geminiviruses that infect a variety of plant species. Although adults are the only truly mobile whitefly morph, and the most important stage for virus transmission, nymphs are capable of acquiring geminiviruses from infected plants and of transmitting virus as adults.[82] Although both sexes are known to vector geminiviruses, females have a reputation as more efficient virus vectors.[82,194] Further, it was noted that A biotype females of *B. tabaci* reared on virus-infected plants were more fecund than those reared on the virus-free plants, suggesting diseased plants may be more suitable as reproductive hosts of this New World biotype. This finding has obvious implications to the primary and secondary spread of geminiviruses in monoculture settings and in naturally infected weeds that are favored as reproductive hosts of the whitefly.[195]

Biotypes of B. tabaci and Differential Transmission

The Discovery of Whitefly Biotypes

It has been shown that the particular whitefly phenotype-genotype and biotype host plant interactions affect the dynamics of virus vector host interactions. These interactions are thought to play a major role in the plant species utilized by different populations, and likewise in the distribution and incidence of geminivirus diseases. The biotype concept addresses this observation and poses the hypothesis that morphologically indistinguishable populations of *B. tabaci* exhibited measurably different biological traits that directly influence virus dissemination. Recent studies have confirmed the widespread occurrence of whitefly biotype phenotypes that vary in host range capabilities, host plant preferences, fecundity, dispersal behavior, and geminivirus transmission efficiency.[12,196,197]

The first recognition of biotypes or 'races' of *B. tabaci* resulted from a study by Bird[196] demonstrating that Jatropha race of *B. tabaci* in Puerto Rico could feed and reproduce only on select Euphorbiaceous hosts, *Jatropha gossypifolia* and *Croton lobatus* (also considered a species of *Jatropha* by some taxonomists), or, essentially a single plant genus. Further, the Jatropha race was the only whitefly vector of the Jatropha mosaic geminivirus (JMV) to and from *J. gossypifolia* and related species within the same plant genus.[12]

Additional evidence for similar host distinction among *B. tabaci* transmission capabilities related directly to host preferences, was forthcoming from studies in Brazil, the documented center of origin of cassava, *Manihot esculenta*. Interestingly, although the monophagous Jatropha race of *B. tabaci* colonizes a Euphorbiaceous host in the Caribbean Basin, it was noted that although the *B. tabaci* from Brazil could not utilize cassava, also in the Euphorbiaceae. But, the Brazilian race readily colonized bean, cotton, soybean, tomato, among other cultivated crops, and many weeds, similar to the host range of the polyphagous Sida race in Puerto Rico.[197]

Further, the introduction of cassava into Africa, *B. tabaci* races or biotypes, indigenous to the continent, not only colonized cassava, but also transmitted the indigenous ACMV to and from cassava.[13] Recent studies of the cassava biotype in Africa have demonstrated its host specificity in that its host range is limited to cassava and a single Solanaceous weed. Further, the cassava biotype differs from an indigenous broad host range population that transmits other geminiviruses to and from a distinct, but broader range of indigenous weed and crop species in the same locale.[198] The basis for the observed differences in host preference behavior is not well studied, but is likely important in the development of biotypes of *B. tabaci*.[78,182,185]

Biotypes and Geminivirus Transmission

Despite the biotype hypothesis, the host range of *B. tabaci* has been summarized as comprising over 300 different genera and/or species. Empirical evidence has only recently been provided that the host range of different *B. tabaci* biotypes (or populations) are quite disparate, ranging from those with extreme or moderate polyphagous capabilities, to those that colonize a single genus or species (monophagous).[80] And, there is some anecdotal evidence that biotypes vary in their ability to readily adapt to, or develop preferences for, 'new' hosts. Another important consideration along these lines are the observed differences in flight behavior of different populations, a behavior which influences the distances these insects characteristically disperse when crowded and/or food becomes unavailable. Regardless of the potential for either extreme or more limited host ranges of the whitefly vector, is notable that the host range phenotypes of most WFT geminiviruses are far less broad than are those of their respective whitefly vector.

The B Biotype

In approximately 1985–86, the exotic B biotype of *B. tabaci* was inadvertently introduced into the US and Caribbean Basin.[181] That the whitefly was different from local whiteflies was recognized by its distinctive host range, the ability to transmit Old and New World geminiviruses,[3,78] an unusually high reproductive capacity, and the inability to mate with the 'A' biotype of *B. tabaci,* a biotype of *B. tabaci* indigenous to the southwestern US.[181,182]

Diseases incited by WFT geminiviruses have also become more prevalent in areas that have remained essentially free of the B biotype. In Pakistan, for example, the cotton leaf curl virus is transmitted by a local biotype of *B. tabaci* that reached epidemic proportion throughout cotton production areas for the last three consecutive years. Hence, increases levels of whitefly vector populations, worldwide is consistent with the notion that similar management and cultural practices imposed on agricultural systems, worldwide, have triggered a similar response and effect by this insect on a global basis.

It is now recognized that a multitude of distinct *B. tabaci* populations exist, worldwide. These populations can be distinguished by differences in host preferences,[199] induction of phytotoxic disorders in plants,[181] esterase electromorphs,[182,185,200,201] DNA sequences,[185,202,203] reproductive incompatibilities, and geminivirus transmission and/or transmission efficiency.[78] Although the B biotype has been shown capable of transmitting all geminiviruses tested to date, differences in transmission efficiencies and host preferences have been demonstrated.[78] However, the role that these particular phenotypic traits play in effecting a transmission event under natural conditions has not been studied.

Whitefly Feeding Behavior

As with other Homopteran insects, feeding is accomplished with stylets which penetrate the sieve tubes of the plant phloem. In larval whiteflies, the flexible stylets usually penetrate through epidermal cells intercellularly. Most prolonged feeding probes target the phloem where penetration is terminated and intracellular feeding is initiated, but periodically, stylets tap into mesophyll cells where intercellular feeding occurs prior to the stylets reaching the phloem.[73] Whiteflies produce saliva that is somewhat analogous to the watery saliva produced by aphids, and a salivary sheath is formed that putatively protects the stylet during feeding activities. Salivary excretions containing degradative enzymes are presumably injected into the host plant to assist with the stylet penetration and feeding processes.[72,73,204] Although the precise mechanism is not known, it is generally accepted that ingestion and subsequent transmission of geminivirus particles occurs during the process of feeding in phloem.

The behavior involved in stylet penetration of the plant host by *B. tabaci* has been studied through the use of electronic feeding monitoring devices, and feeding behaviors are recorded as patterns on an electrical penetration graph (EPG) and are seen as a series of differential voltage profiles, termed wave patterns. Three different wave patterns were associated with whitefly feeding activities.[204] Saw tooth wave forms were always associated with intercellular stylet penetration of the epidermis, mesophyll, and vascular tissue. This wave form was sustained when stylet advancement stopped, the stylet was retracted, and direction of stylet penetration was altered. Empty salivary sheaths were observed in the path of the previous stylet penetration. High-flat wave forms precede by a short burst of sawtooth forms and a transitional wave form (analogous to X waves from aphid studies) were correlated with ingestion of contents from the phloem sieve elements. Another wave form observed periodically was associated with ingestion, but feeding was not from vascular tissues. Interestingly, the B biotype was able to reach the phloem and initiate feeding in the phloem in an average of 16 minutes,[204] almost twice as fast as previously estimated. This observation corroborates results from several whitefly transmission studies in which a 15 minute acquisition access period (AAP) was adequate to achieve geminivirus transmission, albeit at low efficiency.[1,12,17,69,70,78,205,206,207,208,209]

Whitefly-Mediated Geminivirus Transmission

Geminiviruses are thought to be transmitted as intact virions or nucleoproteins, and transmission is considered to involve a circulative process, or one in which virions must pass through various barriers within the vector prior to reaching an appropriate, but as yet unknown location in the mouth parts, from where particles then are subsequently released or 'transmitted' during the feeding process. Viral DNA can be detected in the whitefly vector,[207,209,210,211,212] and it is estimated that approximately 600 million TYLCV genomes (1 ng of DNA) can be taken up by the whitefly vector following feeding on an infected plant.[209] Interestingly, geminivirus DNA was also detected by DNA hybridization in the whitefly, *Trialeurodes vaporariorum* (West.),[210] which has a host range similar to that of *B. tabaci,* but is not a geminivirus vector.

Recent evidence indicates that whiteflies ingest from the plant, not only virions, but dsDNA and possibly unencapsidated ssDNA forms as well.[209,213] These data suggest that the process by which whiteflies initially ingest the contents of the sap containing viral DNA and/or virions relies solely on the passage of the sap into the gut via the mouth parts. Because the vector-transmissible form of the virus, as well as free or complexed viral DNA forms that are as yet unencapsidated, are also present at the same time in the cells upon which whiteflies feed, all forms are conceivably taken up by the vector, but those unprotected by encapsidation are likely to be degraded in the insect gut shortly thereafter. Size selection for virions, naked viral DNA forms, or other nucleoprotein complexes would be below the physical limitations of the diameter of the stylet, estimated to be about 0.66 μm.[214]

Transmission Process.

Following virus ingestion or uptake, geminiviruses have a latent period of approximately 6–12 hours in the vector, and during this period, although viral DNA can be detected in the whitefly, inoculation access feeding on a host plant will not result in virus transmission. The latent period is the amount of time required for virions to be routed through barriers in the whitefly vector mouthparts, gut, and/or hemolymph prior to reaching the site where virions are sequestered until subsequent transmission to a plant host during whitefly feeding. Neither identification of such putative site(s), nor an understanding of the cellular basis for the specificity between geminiviruses and the whitefly vector have been accomplished. However, one possibility is that virions become associated with the salivary glands of the insect during some critical point in the acquisition and/or transmission processes, as has been shown for the phloem-restricted luteoviruses that have aphid vectors.[215,216] This type of relationship has been categorized as persistent, or one characterized by a minimum AAP of several minutes to hours, a latent period of several hours, and an inoculation access period (IAP) of minutes to hours. Once virus acquisition has been accomplished, the vector can continue to transmit virions for days or throughout the entire life of the vector.

The transmission process has been monitored over time for several geminiviruses and the respective whitefly biotype using bioassay, DNA hybridization, and polymerase chain reaction assays to detect viral DNA in whitefly adults.[82,207,209,210,212] Using a probe for TYLCV and DNA hybridization, only viral monomer-size ssDNA was detected in the whitefly vector, despite the detection of both ss and ds viral DNA in plant sources.[211] In a follow-up study in which AAP varied from 30 min to 12 hr, viral DNA was detected in some insects assayed at all AAP's tested, but viral DNA was detected with greater frequency within a cohort as AAP was increased over time. By 90 min, virus DNA was detected in all whiteflies that had been given the opportunity to feed on TYLCV-infected plants. The amount of DNA (about 600 million genomes equals 1 ng of

DNA) detected per whitefly also increased with increasing AAP, and ranged from 60–400 million genomes per insect, after 1 hr AAP, to 20–100 million genomes in nearly 60% of whiteflies after an 8 hr AAP. Over 90% of the insects had acquired 100–600 million genomes after a 12 hr AAP.[209] After a maximum AAP of 48 hr, viral DNA was detected by autoradiography in about half of the whiteflies assayed over a 13 day period. Virus DNA levels in the whitefly began to decrease 2–12 hr after acquisition feeding was initiated, followed by an increase 6–36 hr later,[209] in conjunction with concurrent increases in virus levels in the plant.[217]

In contrast, results from a study of temporal transmission patterns of the New World bipartite virus, SqLCV, viral DNA was detected in over half of the whiteflies tested after a 48 hr AAP.[210] In both studies, irrespective of increasing AAP, the amount of virus ssDNA that was detected in the whitefly did not exceed 600 million genomes or 5 ng of viral DNA.[209,210,212]

Virus retention.

Once acquired, geminiviruses are theoretically transmitted throughout the life of the vector, or for approximately 10–15 days. For whitefly-mediated TYLCV-Is transmission, a phenomenon termed periodic transmission was coined to describe the inability of the whitefly vector to acquire additional virus until the charge obtained from the initial AAP had been completely exhausted. The ability to transmit virus could be restored following a second AAP, only after first loosing the ability to transmit about 10 day after the initial AAP.[206] The mechanism underlying the inability of whiteflies to transmit TYLCV in a continuous manner, despite provision of a supplemental AAP prior to complete loss of the initial virus charge is not known. Further, it is unclear how whiteflies can be relieved of the acquired virus charge when, following a successful acquisition, whiteflies are thought capable of continuous or at least intermittent transmission for the life of the vector. In any case, these results, collectively, suggest that whiteflies acquire a maximum virus load, given a saturation-level AAP on virus-infected plants, and these observations may further be reflective of a steady-state turnover in incoming virions that are either degraded or somehow not destined for transmission, versus those that are retained by the vector and/or ultimately sequestered for potential transmission events. Whether these parameters vary with whitefly biotype and/or virus is not well documented, but most evidence to date supports the notion that the process, although well orchestrated, is highly complex, and as yet poorly understood.

Virus-Vector Specificity and Mechanism of Virus Transmission

Early evidence of the role of the capsid protein in insect-mediated transmission for other plant virus groups stimulated speculation for direct or indirect involvement of the coat protein in whitefly transmission of geminiviruses. Comparison of the deduced amino acid sequences of capsid proteins of geminiviruses indicate a high degree of similarity (approximately 84–97% at the amino acid sequence level) between coat proteins of WFT geminiviruses, but a low degree of similarity (80%) between the capsid proteins of members of both leafhopper subgroups. That members of the WFT subgroup share a common vector species whereas, in general, members of Subgroups I and II have separate vector species, suggests a conserved mechanism of specificity for the WFT viruses.

Detailed sequence comparisons of putative viral genes and gene products of the WFT subgroup indicates that the capsid gene and its gene product are the most conserved among Subgroup III genes and viral-encoded proteins, respectively. Inferred phylogenetic relationships based on sequence comparisons of the capsid genes and vector relationships are illustrated in Figure 3.[165]

This high degree of gene conservation, and the fact that all WFT geminiviruses, described thus far, are transmitted exclusively by a single whitefly vector, *B. tabaci*,[3,31,78,101,201] suggest a precise role for the coat protein in vector-mediated geminivirus transmission. Coat protein mutants of bipartite subgroup III geminiviruses that do not produce a capsid are not transmissible by the whitefly vector.[107] Replacement of the coat protein gene of ACMV, a WFT subgroup III virus, with that of BCTV, a non-whitefly transmissible (leafhopper transmitted) subgroup II geminivirus, resulted in a chimeric virus that was transmissible by the leafhopper vector.[108] This latter experiment provides compelling evidence for the essential role of the capsid protein in the transmission process.

The basis for geminivirus-vector specificity conceivably involves a mechanism in which the viral capsid functions directly in virion-whitefly interactions, or one in which the capsid plays an indirect role through interactions with a transmission factor-like molecule (TFM), the latter mechanism typically invoked for noncirculative plant viruses. Regardless, in either scenario, epitopes exposed on the surface of virions possibly interact directly with either a TFM or with specialized sites in the insect vector during various stages of the feeding and transmission processes.[101,109]

Cellular and Molecular Interactions Between Virus and Vector

Although the precise cellular and molecular mechanisms involved in the process of geminivirus transmission have not been elucidated, recent studies have provided the first road map of the internal and external anatomy of the whitefly vector and its mouthparts, obvious prerequisites to detailed mechanistic studies at the cellular level.

Whitefly mouthparts, composed of the labrum, labium, and stylet bundle, are similar to those described for other Hemipteran phloem specializing insects that locate the vascular bundles and ingest phloem sap from which nutrients are obtained. The stylet bundle is composed of two pairs of mandibular and two pairs of maxillary stylets. The mandibular pair of stylets function in piercing and cutting of leaf tissues and probably help anchor the stylets during feeding, whereas the maxillary stylets are interlocked to form two compartments: the food canal and the salivary canal. Scanning electron micrograph's indicate that the salivary canal (diameter 0.21 mm) is contained almost entirely within one stylet, while the food canal (0.66 mm) is centrally located and formed by apposition of both stylets. Both canals are more than adequate to permit passage of geminivirus nucleoproteins of approximately 20 X 30 nm (0.020 X 0.030 mm) in size. The labium of females are longer than males, or about 300 and 215 mm long, respectively.[165,214]

In whiteflies, the stylets are flexible, and enter the plant by penetrating intercellularly. Most feeding probes terminate in the phloem.[73,204,218] Phloem contents is likely acquired through the food canal, whereas the saliva in injected into the plant through the salivary canal. Based on mouthpart morphology and ultrastructural evidence of stylet penetration of plant tissues,[218] it has been proposed that only the maxillary stylets penetrate the vascular elements during sap uptake.[72,165,214] This is consistent with the 'mandibular stylet ahead' hypothesis that supports the evolution of single stylet innervation,[72] and the exclusive use of mandibular stylets for penetration of epidermal and parenchyma cells and as a guide and support the more delicate maxillary stylets that function in uptake of sap from the phloem.

The alimentary system of whiteflies is unique relative to other Hemipteran insects in that it lacks a peritrophic membrane. The midgut of *B. tabaci* is looped positioning the anterior and posterior extremities such that they come into contact through adjacent basal lamina, making the opposing epithelial membranes of both extremity contiguous. The esophagus spans the thorax and is connected to the midgut and paired salivary glands (J. Cicero, pers. comm.).[219]

A filter device is located near the distal portion of the hindgut where it is found in close apposition with the proximal portions of the midgut.[214,220] The filter device may be involved in diffusion of excess water and soluble carbohydrates directly from the anterior alimentary canal into the posterior portion of the intestine, while amino acids, proteins and lipids are retained to be digested in the more anterior portions of the alimentary canal (J. Cicero, per. comm.).[221] During feeding, fluids likely move down the esophagus, collect in a crop-like structure, pass into the lumen of the filter chamber, and are diverted into the midgut which serves as a temporary reservoir. As water is eliminated through the filter chamber, fluids are allowed to pass into the hindgut.[220] Preliminary studies indicate that the basement membrane of the hindgut is extremely narrow, and the cells contain an abundance of mitochondria and endoplasmic reticulum, suggestive of a high degree of metabolic activity and of potential sites of transport between the hind gut and the hemocoel.[222] This knowledge of the whitefly mouthparts and internal structures will be valuable in the tracking of virions during the ingestion, acquisition, and transmission processes.

Basis for Transmission Specificity

A potentially applicable example and mechanism of virus-vector specificity has been advanced for the aphid-transmitted luteovirus, barley yellow dwarf virus (BYDV), the only well-studied plant virus-vector system involving a persistently transmitted virus and an Homopteran (Hemipteran) vector. However, in the yellow dwarf system, a high degree of vector specificity occurs between vector species and virus strain. In contrast to the geminivirus-whitefly system which is characterized by a universal vector and many viral quasi-species, several strains of BYDV have either a single aphid species as vector, and one virus strain can be transmitted by more than one aphid species. This system[223] has been the foundation for detailed fundamental studies in which aphid vector-virus specificity involving transmembrane passage of virions by coated vesicle-mediated transport processes have been elucidated in part, during the past thirty or more years.[216] In this virus-vector interaction, virions are acquired by the vector through feeding in the phloem, pass via the stylets into the midgut, into the hindgut and through the hindgut epithelium, unimpeded to the hemocoel. Virions of all virus strains tested have been shown to pass through the hindgut membrane of both vector and non-vector strains. Some inherent selectivity in exiting the hindgut is postulated because virions have been shown to cross this barrier by classical receptor-mediated endocytosis.[224]

BYDV particles have been observed in the hemocoel of the aphid and at the external and internal surfaces of the accessory salivary gland (ASG). The function of the ASG is production of salivary excretions that are required for successful feeding in the phloem, and thus from a virus perspective is a likely candidate for a site of virion sequestration. Virus-vector specificity in the BYDV-aphid system is correlated with selectivity of the basal lamina membrane of the accessory salivary glands that either accommodates or disallows passage of virions into the cytoplasm of the glandular epithelial cells.[216] Virions that crossed the plasma membrane of the known strain of vector were always transmitted, whereas virions that did not cross the basal lamina of non-vector strains were not transmitted. These and other more detailed studies suggest the involvement of the basal lamina as a transmission barrier, and implicate virus-specific capsid-glycoprotein interactions in the regulation of membrane penetration by virions.[215,216,225]

In light of these observations, it is of interest to know what operational mechanisms are involved in whitefly-mediated transmission of geminiviruses. Also unknown is whether virions utilize an additional viral encoded gene product(s) or helper factor. Further, the identify of putative virus-specific whitefly cellular factors has not been elucidated in *B. tabaci*-geminivirus systems. Detailed studies are now needed to delineate the precise role of the capsid protein in whitefly-mediated transmission, and the underlying cellular and molecular mechanisms involved in this apparently specialized interaction.

Evolution of Co-Localized Sites in Plants for Virions and Vector Feeding

Whitefly Feeding Sites and Virus Ingestion

Recent molecular studies of 18S rDNA indicate that whiteflies share a common ancestor with aphids and scales, and these insects are also vectors of plant viruses in subtropical/tropical climes.[226,227] Studies of the feeding mechanisms used by phytophagous insects indicate that whiteflies, aphids, and scales probe intercellularly using stylets to ingest nutrients from phloem sieve elements, and produce a stylet sheath.[72] This type of feeding strategy in insects has been linked to the evolution of lignophytes that were the first plants to have specialized vascular tissues, ultimately giving rise to Angiosperms in the lower Cretaceous. Feeding on phloem sap is thought to be a reflection of the subsequent advanced photosynthetic processes in chlorophyllous Angiosperms, and phytophagous Hemipterans are believed to be the most recently evolved among insects that specialize in phloem feeding.[226,227]

The evolution of a conserved specificity between plant geminiviruses and a single whitefly vector species evidently occurred as a result of frequent interactions over time between *B. tabaci* and the plants that served as hosts of WFT geminiviruses. Indeed, it is no coincidence that geminiviruses infect the phloem, and that temporal expression of coat protein in the phloem and mesophyll appears to be regulated by a virus gene that either represses or derepresses transcription of the coat protein gene in the respective tissues at different times during the infection cycle.[42] It is no surprise, then, that interactions between virions, the whitefly vector, and the host plant that are essential to successful vector-mediated virus transmission have evolved into a highly specialized system that facilitates the effective mobilization of geminiviruses from host to host.

In light of these observations, some important questions can be raised regarding the evolution of this biological system, and that if answered, could lead to valid predictions about relevant developmental processes in the whitefly vector and geminivirus plant hosts. For example, did geminiviruses evolve first as viruses in the phloem, and later developed, but only in some cases, the ability to utilize mesophyll cells as well? Why is the mesophyll targeted by some geminiviruses but apparently not by others? Are virions encapsidated in one or both cell types, or are they assembled only in the phloem but are capable of moving between cell types only in certain plant hosts? Can geminiviruses be acquired from both the mesophyll and the phloem, or is the ability to inhabit the mesophyll a more recent development, and an *in planta* location that does directly serve as a source of virions during whitefly-mediated acquisition and transmission? Could the ability to utilize both mesophyll and phloem eventually lead to the development of a distinct virus group that differs from extant geminiviruses?

Recall that whiteflies specialize in phloem-feeding, and are thought to feed only periodically in mesophyll cells (as adults), but only by 'accident' while probing in an attempt to reach the phloem.[73] However, there is some evidence that stylet tracks of the immature stages of whiteflies are more frequently found in mesophyll than are those of adults. In either case, acquisition and transmission by the adult whitefly is presumed the most important mode of virus dispersal, emphasizing the likelihood of mere chance acquisition of those virions not present in the phloem tissues. What advantage does presence in the mesophyll confer to extant geminiviruses? It is possible that in the event of a drastic change in host localization and/or vector transmission modes, ancillary presence in more than one host tissue might facilitate alternative opportunities for different modes of transmission or rescue from the plant? What those bottlenecks might be, and under what conditions such a scenario would evolve is limited to pure speculation at this time.

Virus Localization in Plants and Transmission

That geminiviruses need dividing cells for replication has been postulated from protoplast studies with ACMV.[184,228] In this scenario, meristematic tissues would be the target of replication and the associated viral activities. Interestingly, there is no evidence for accumulation of viral DNA in dividing cells *in planta* based on DNA hybridization and localization studies. Instead, viral DNA was detected in already differentiated phloem.[65,66] Recent studies that link at least one of the functions of the AL2 gene product to regulation of AR1 and BR1 expression in a tissue-specific manner,[229] and the putative function of AL3 in regulation of the cell cycle (Bisaro, pers. comm.) provide compelling evidence for temporal and spatial regulation of rightward gene expression during an infection cycle leading to the production of virions in the precise location when and where the whitefly vector feeds.

The processes and mechanisms involved in the specific targeting of assembled virions in infected cells and tissues are not completely understood. Delineation of these factors will clarify the relationship between tissue-specific viral gene expression, the localization of virions to specific sites within the host plant, and the ability of the whitefly vector to effectively mobilized virions to a host plant. Still unexplained is the propensity of geminiviruses for certain tissues of their host plants, and further, that some viruses are present in several tissue and cell types in several different hosts examined, whereas others are highly restricted to certain sites in all host species studied. The encapsidation of viral ssDNA several distinct cell and/or tissue types, accompanied by the requirement for long-distance vector-mediated dispersal must be understood initially in the context of the site(s) from which assembled virions are subsequently acquired by the whitefly vector. What are the implications of the differential tissue localization observed for WFT geminiviruses in light of a knowledge that the insect vector has evolved a feeding behavior that restricts feeding to the phloem of the plant host. Is extra-phloem localization a remnant of the past, or insurance for the future? Precisely where whiteflies feed with respect to host tissues and to the physical mechanics of viral uptake are poorly understood, but the convergent evolution of the interaction between the feeding behavior of the vector and the specific host sites occupied by potentially transmissible virions is of utmost importance to the sustainability of the Subgroup III geminiviruses.

The lack of fossil records or other means by which to study the archeovirology of the geminiviruses precludes direct assessment of progenitor viruses and their associations with plant and other hosts, and with their whitefly vector. Of interest, never-the-less are the means for exploring the archeovirology of geminiviruses. In this light, do present-day geminiviruses that are limited to the phloem represent the more primitive virus, solely able to function in those tissues, as compared to those geminiviruses that routinely inhabit the phloem and mesophyll (and in at least one case, the epidermal cells)? What are those determinants that confer the propensity for physical location in the host plant? Further, how did these determinants evolve in geminiviruses to reach the present state of pathogenesis, and how do such putatively expanded viral functions restrict or promote the evolution of the viruses?

Vector transmission appears to be a highly specific process involving not only localization of virions in the appropriate host tissues and site within those tissues for uptake by the whitefly, but putatively highly specific virus-vector interactions as well. Are selection factors at work then at the level of virus-host plant interactions that may provide new alternatives to the virus for mobilization by a different vector? If so, how successful will geminiviruses be if the specificity between the capsid and whitefly is either reduced or eliminated. Or, does the basis for the apparent specificity at the geminivirus-vector level, reside instead in a widely available divergent mechanism present also in other insects that feed in the mesophyll and phloem, a mechanism that in time, may prove important in the evolution of a mechanism to permit transmission by additional or other insect vectors, and hence to the survival of geminiviruses as successful viral pathogens?

TOWARD MOLECULAR EPIDEMIOLOGICAL PERSPECTIVES

Agroecosystems And The Environment

An important feature of monoculture production in subtropical and mild temperate zone climes is the widespread practice of continuous cultivation of irrigated crops with rare or short time spans during which crop free periods are observed. Other features include the cultivation of crop cultivars with highly uniform characteristics centered around hybrids or segregated genetic pools, and the demand for specific horticultural characteristics for particular markets. Many of these cultivars have not retained the capacities inherent in wild germplasm that confer tolerance or resistance to insects and pathogens. In addition, the associated high inputs of insecticides, fertilizers and other hard chemicals negatively impact the insect flora. Contemporary agricultural systems have great potential for producing abundant, high quality food on a year round basis, and in general, these goals have been achieved despite unanticipated or expected potential shortfalls. Amidst our successes also lurk numerous underlying possibilities for alternative consequences of the imbalances imposed upon the system. Among these are pathogen and insect populations that will respond to selection pressures imposed by agricultural practices, emerging as increasingly virulent viral diseases and more frequent infestations of arthropod vectors and pests.

The cultivation of crops under present conditions encourages a high disease incidence and virulent virus populations due to continuous exposure to host-imposed constraints, high density vector populations, and environmental stresses. The high density plantings of virus host plant species in close proximity to other hosts fosters opportunities for contact between virus populations and non-virus host species, owing to the typical host searching/tasting behavior and polyphagy of the whitefly vector. Opportunities are likewise increased for co-infection by two or more geminiviruses, for higher mutation rates, and intermolecular recombination and pseudorecombination between compatible genomic components. The number of apparently distinct whitefly-transmitted (WFT) geminiviruses now documented in crop plants, worldwide, has reached an unprecedented high, and there is a general lack of information concerning phenotypic and genotype differences and similarities among this subgroup of plant viruses. A critical examination of the viral genotypes that are now predominant as pathogens, and the underlying factors that may have influenced these changes requires the application of molecular epidemiological tools, now within reach for the first time.

Genetic Variability And Geminivirus Quasi-Species Concept

The inherent properties of plant viruses are decisive determinants of the outcome of interactions between the virus and host plant, and the virus and insect vector.[9] This inherent concept is recognized by the frequent reference to genotype and phenotypic characters used as working definitions of plant viruses, further supported by observations that some viruses are seemingly stable over time, while others appear more dynamic. The term viral quasi-species has been introduced to describe these selective interactions and the output of a predominant genotype/phenotype, relative to specific selection forces which remain for the most part, unidentified at present for most viral quasi-species. The recent ability to directly evaluate genotypes and corresponding viral phenotypic markers has led to the conclusion that virus isolates are compositionally dynamic genomes, collectively capable of potentially endless variability that can be exploited under a wide array of subtle or extreme selection pressures. More definitively then, the concept of quasi-species most aptly define the inherent plasticity of virus populations.[230,231]

The initial indication of a shift from one predominant genotypic condition to another may be recognizable only as a differential symptomatological phenotype in a plant host, or as a change in host range. It follows that the first step in effectively reducing the impact of a pathogen is to know the identity and distribution of the virus pathogen of interest. Traditionally, plant virology has emphasized the interplay of the viral phenotype with environmental parameters in discovering viral variants and subsequently defining their properties. This approach requires defining modes of virus transmission, the interactions between plant hosts and virus vectors, and ultimately mapping the distribution of virus quasi-species in the context of the agroecosystems.

With genetic variation inherent in virus populations, the genotype and resulting phenotype of the predominant population or quasi species in a host can evolve or change dramatically in a short period of time.[82,230] The selection pressures that drive such shifts are poorly studied in mechanistic terms for plant viruses. In contrast, there are several documented interactions between animal virus-host interactions, for example, mammalian host immune systems that select for new, more fit (non detectable) viral genotypes, as evidenced by frequent shifts to new or different serogroups in order to circumvent host defenses.[6] In plant viruses, selection of new strains or predominant genotypes from the quasi-species pool can be exerted by various factors that include among others, host plant genotype, environment, specific virus resistance genes, a dependency on specific arthropod, nematode or other biotic vector for transmission, vertical transmission constraints, and association with a helper virus, molecule, or other dynamic facilitating factors.

Selection of Genotypes

As noted, it is not uncommon to discover mixtures of WFT geminiviruses in the same plant, as well as multiple infections of related and unrelated viruses. Also, documented is the ability of the whitefly vector to transmit more than one geminivirus during the same inoculation access feed, resulting in delivery of more than one geminivirus presumably simultaneously to the same host plant. The apparent explosion in the number of distinct WFT geminiviruses in several locations throughout the world in conjunction with unprecedented increases in whitefly vector population levels has led to the suggestion that the geminiviruses are rapidly mutating in response to new selection pressures. In this scenario, the vector and host plant passage are likely candidates for exertion of such pressures.

Genotypic Variation and Mutation

DNA-containing viruses are thought to have mutation rates comparable to their eukaryotic host plants, approximately 10^6–10^9 per round of replication, compared to mutation rates of some RNA viruses estimated as high as 10^3–10^5 due to substantial infidelity of RNA polymerase.[232] In several well-studied systems, there is evidence to suggest that genetic drift operates at particularly susceptible sites in the genome, but to date, there is little definitive evidence for specific mechanisms of mutation in wild type or laboratory populations of geminiviruses.[233] Although genetic drift is presumed, certain observations described throughout this writing may suggest favored means for effecting genetic change among the group. Based on manifestation of disease, the alterations in host plant-quasi-species interactions can be additive or synergistic with respect to symptom phenotype, resulting either in greater or less severe virus virulence, or may have minimal (or neutral) consequences when one quasi-species out-competes another. The displacement of one quasi-species by another can be envisioned through several mechanisms including more effective use of host machinery (faster or greater rates of reproduction), more compatible movement factors, or selection through greater vector efficiency. The selection forces, that act on wild type geminiviruses and give rise to a consistent or stable genotype/phenotype, and how new quasi species are erected in nature, have not yet been addressed.

Inter- and Intra-Molecular Interactions

Alternatively, when viruses occur in mixtures, it is more than feasible that new quasi-species are selected through recombination, pseudorecombination, and transencapsidation, or other intermolecular interactions, as has been discussed in a recent review on the subject.[233] The hypothesis that geminivirus intergenic regions are reflective of the potential for specific intermolecular interactions with other viruses, creates opportunities for epidemiological studies of diseases incited by mixtures of geminiviruses. Can the potential for virus interactions in mixed infections, in part, explain the seemingly sudden explosion of different phenotypes in Mexico? This phenomenon itself leaves those who have investigated the situation over the last decade to postulate on the existence of a high degree of variability, and on the apparent rapid evolution of geminiviral pathogens in Mexico and the Caribbean Basin countries.

In contrast, although lacking in breadth in terms of historical observations, there is precedent for several putatively stable geminivirus genotypes of historical importance, albeit the time span is short. For example, there is no evidence for drastic changes in genotypes (based on phenotype) of BGMV throughout the Caribbean Basin, TYLCV (prototype) in tomato in the Middle East, and ACMV in Africa. Whether the narrow host ranges of these viruses is an indication of successful viral pathogens, and hence the absence of measurable change over a recent time-frame, has not been investigated. The inherent stability among geminiviral genomes have been demonstrated by experimental attempts to create laboratory variants by making critical mutations in essential genes. In these examples, geminiviruses consistently reverted to either the original genome or genome size.[4,193,233,234] The conditions that favor this inherent stability of and within viral genomes are not known. However, in the latter case, it is safe to speculate that plasticity of the genome can only occur in regions of the genome not involved in gene function critical to short term survival of the quasi-species.

The recent increase in disease incidence worldwide, however, can not be explained exclusively in terms of altered viral genotypes as a result of virus-virus and virus-host interactions without a consideration of other relevant selection forces. Also important to consider are the recent and unprecedented pressures exerted by the whitefly vector through whitefly-host plant and whitefly-virus interactions.

Virus/Vector/Host-Interactions and Epidemiology

The dynamics of selection of a predominant quasi-species involve complex interactions between the virus, vector, and host plant. Total transmission rates from a host depend on the integral rate of transmission over the duration of the infection.[6] Thus, whether there is either strictly seasonal or year-round availability of a suitable vector is critical to the ultimate selection of a particular quasi-species relative to successful interactions with the host, and ultimately to survival of the virus. Until recently, the geminiviruses were pathogens present in a few cultivated crop species, and transmitted by indigenous whitefly vectors that reached high densities typically only in association with monoculture crops (fewer and smaller areas were devoted to monoculture until quite recently), and during the dry season in the arid-subtropics. Accordingly, epidemiological theory would predict that these viruses would be moderately virulent because of a dependence on a seasonal vector that reaches adequate levels to achieve virus dispersal to either weed or crop plant hosts, whose presence and densities were in turn, also dependent upon seasonal rainfall.[6,82]

Golden Mosaic Phenotypes

Viruses that exhibit lower rates of replication do not kill the host plant as rapidly as faster replicating genotypes, which permits a trade-off between lower host mortality rates and greater probability of transmission over time. This could facilitate host and virus survival over long periods of time during which the vector is either entirely absent or present but at low densities. Further, localization of available virus at numerous host sites, i. e. a variety of vascular tissues, and/or in mesophyll and epidermis, would be another strategy toward increased probabilities of ingestion by a scarce or seasonal vector. Although purely conjecture, these characteristics are reminiscent of golden mosaic Subgroup III phenotypes. In these phenotypes, is there a correlation between the intensity of the bright yellow mosaic and the need to attract a whitefly vector, routinely present either at low densities and/or with seasonally adequate levels? It may be of interest to note that the majority of Subgroup III viruses documented worldwide, were more frequently golden mosaic symptom phenotypes, as opposed to the more virulent leaf curl/mosaic phenotypes of more recent occurrence. Taking these observations into consideration, the conditions under which the putatively once rare leaf curl phenotypes occurred could be examined to test the utility of these concepts in predicting outbreaks and consequences of geminivirus diseases.

Leaf Curl/Mosaic Phenotypes

The converse of this scenario would be the phloem-limited nature of the more virulent leaf curl geminiviruses that predominate or emerge under high vector densities, which offset the probability of fewer opportunities of ingestion by the vector with that of greater rates of transmission (i. e. higher virus titer). A shorter duration of infection before the host plant is killed or incapacitated (to the degree that replication is limited), is in sharp contrast to the alternative strategy in which viruses have a longer duration in the host due to lower rates of replication. It follows that the more virulent, faster replicating viruses, possibly localized in limited cells or tissues, should be selected for under conditions of high vector densities and a requirement for less dependency on seasonal availability.[6]

Would one predict then, that under such circumstances, highly virulent quasi-species would predominate, and continue to be selected for as long as these conditions remained unchanged? Is this similar to the situation that has occurred with the rise of vector densities, nearly worldwide, and the subsequent emergence of more widespread disease incidence and the emergence of several highly virulent viruses in those locates? Consider next, the effects of construction of irrigation systems that permit cultivation of at least two crops instead of one per year, and the introduction of crop varieties that perform readily under previously prohibitive conditions of heat or cold, allowing for extension of growing seasons. Then, superimpose the selection of insecticide resistance in the whitefly vector due to high-input farming practices, instigated to produce sufficient food for the world population, with the apparent ability of the whitefly to disperse unimpeded across irrigated, arid and dry-subtropical valleys. Are we viewing the present situation described as the emergence of the subgroup III geminiviruses as unprecedented pathogens in numerous cropping systems, worldwide? Although oversimplified as stated here, and only indicative of the extremes, these principles may now be applied to testable hypotheses to define the epidemiology of diseases incited by WFT geminiviruses.

Recent Studies: Molecular Epidemiology of Subgroup III Geminiviruses

A better understanding of geminivirus-whitefly vector-host complexes will facilitate studies concerning the affects of variability, plasticity, and stability of the respective genotypes on the development of disease epidemics. Ultimately, these data should facilitate design of disease control strategies that utilize a knowledge of the most predominant virus quasi-species and vector biotype(s) in defined agroecosystems. In the following section, the results of several studies along these lines are reviewed in an attempt to examine the initial frameworks of molecular epidemiological perspectives of the Subgroup III geminiviruses.

Evolution of African Cassava Mosaic Virus Strains

The ACMV is prevalent in cassava throughout the African continent.[235,236] Examination of the variability of geographically representative isolates using a panel of monoclonal antibodies (Mab's) to ACMV and another WFT geminivirus of cassava from India (ICMV)[237] indicated at least three different stains of the ACMV were present on the African continent, and that ICMV is distinct from ACMV strains.[101,114,112,237] Patterns of variation in seventeen epitopes of the capsid protein were examined for 87 isolates from ten African countries, permitting an estimate of the degree of variability, the distribution of isolates for which typed profiles were obtained, and an investigation of genetic drift. Epitope patterns revealed three distinct groupings of isolates from Africa with overlapping or shared epitopes, termed A, B, and C. All isolates reacted to Mab SCR 20, the putatively universal Mab thought to react to an epitope present on all WFT geminiviruses, with a postulated role in vector specificity.[101]

In DNA-DNA hybridization assays with cassava isolates from the three subgroups, reactions with A or B component probes indicated that the B components varied somewhat between isolates, whereas the A components cross-hybridized. From these results, it was postulated that the observed low degree of capsid epitope variability was indicative of strains of ACMV and not of distinct viruses, and that strain differences were probably related to selection by interactions with the host plant. The occurrence of three extant strains of ACMV is hypothesized to reflect temporal differences in host-virus interactions brought about as a result of infection of cassava by an indigenous virus at three different periods during the 16th to 18th centuries, when cassava was introduced into Africa from South America. Cassava reached the Congo region and moved westward in the 16th and 17th century (ACMV-West African strains), the Cape of Good Hope and to the East African coast in the 18th century (East African strains), and Reunion or Mauritius eastward to India and Sri Lanka in the 18th century (Indian cassava mosaic virus).[101]

The longevity of interactions between a host and it virus seems an important factor when considering evolution from non-host to host status. In this particular scenario however, the impact of longevity of virus-host interactions seems negligible since the virus probably infected cassava shortly after introduction of the plant into each region. Only recently has cassava germplasm been mobilized between regions, potentially permitting the introduction of new strains, and the homogenization into one or a few predominant strains per region. The detection of several extant geographically distinct strains suggests that there are different, or additional contributing factors that may be considered.

Although epitope profiles do indeed indicate geographic partitioning of ACMV strains in Africa, an alternative explanation for strain variation may instead involve the notion that variants already somewhat evolved from the same quasi-species (experiencing expected genetic drift) were present in the three different regions into which cassava was introduced over a two hundred year time-span. This scenario could be envisioned if strains are thought of as having arisen initially from a common ancestor, millennia prior to the introduction of cassava to Africa, followed by isolation of isolates due to creation of new geographic barriers resulting from continental drift. The variability between strains is not necessarily related causally to the coincidental introduction of cassava, but could be linked instead to present-day physical barriers on the continent that made travel by sea to different ports easier than was crossing by land. If humans did not readily pass the geographic barriers, it is not likely that viruses or whitefly vectors could either, corroborating the notion that the variants of ACMV were possibly present at each site, but already geographically isolated and somewhat genetically variable from one another when cassava arrived on the continent. Appropriate detailed molecular analysis will permit an estimation of mutation rates of geminivirus genomes,[163] and also make possible hypotheses about additional mechanisms by which ACMV strain differences could have arisen.

Of obvious importance to this question, then, is identification of the putative indigenous plant host that served as the source of the ancestor ACMV in Africa. Because the predicted center of origin of cassava is South America, and there are no geminiviruses in South America known to infect the species, the origin of the virus cannot have involved cassava germplasm-mediated introduction of an exotic virus from South America to Africa. Thus, the source of the virus to introduced cassava was likely an endemic plant species other than cultivated *Manihot*, and most certainly a weed host as well. It follows that the progenitor isolate probably experienced a certain amount of genetic drift in association with cassava, that may not have occurred concurrently during constant association with the host species harboring the indigenous virus, now known as ACMV. Thus, it is possible that an ACMV relative exists in an extant weed species in Africa and from which, if identified, the molecular variation of the parental quasi-species and associated phenotype(s) can be identified. In contrast, based on epitope profiles, the geminivirus isolates from cassava in India are distinct from those from African isolates, despite the ability of both viruses to infect the same host plant. Owing to present geographical barriers in Africa and the Indian Subcontinent, and the resulting potential isolation of cassava plantings and their whitefly vector in those agroecosytems, it may not be surprising to find additional genotypes on the continent that should be included in regional efforts devoted to plant breeding or genetic engineering of resistant varieties. The effect that propagation of cuttings infected with ACMV has had on the genotypes and dispersal of quasi-species during the last several decades has yet to been examined in the context.

Possible Evidence for a Bean Golden Mosaic Virus (BGMV) Quasi-species

Examination of DNA sequences for several BGMV-like isolates from the Caribbean Basin indicates that isolates which produce golden mosaic symptom phenotypes in common bean are genetically similar to one another.[166,172,238] Isolates examined thus far have been collected from Cuba, Dominican Republic, Guatemala, Jamaica, and Florida and Puerto Rico in the US, and are either very similar or identical. Those isolates for which host range studies have been conducted have a narrow host range, and are like that described for the type isolate of BGMV from Puerto Rico.[239] The low degree of genetic variability among all these isolates suggests that they collectively represent a distinct virus species (quasi-species), or possibly strains of the BGMV that have evolved to a stable closely-related quasi-species. Forthcoming data at the molecular level will be important in clarifying this situation.

In contrast, an isolate that induces similar golden mosaic symptoms in common bean from Brazil is genetically distinct from the type BGMV and strains thereof in the Caribbean Basin.[240] Unfortunately, the virus has been erroneously designated BGMV-BZ, suggesting that it is a strain of the type BGMV when in fact there is strong evidence that it is a distinct virus causing a similar symptom phenotype. As with other transitions, the nomenclature of the WFT geminiviruses will take some time to rectify, and additional examples of redundancy in species names will become apparent as more complete characterization of the many currently uncharacterized viruses is accomplished. Several other isolates that infect bean, but cause calico symptoms, for example, bean calico mosaic virus (BCMoV) from northern Mexico,[241] or severe dwarfing as observed with the Colombian bean dwarf mosaic virus (BDMV), appear to constitute distinct viruses.[122,242] In depth knowledge of such differences and similarities have relevance to plant breeding programs. Long recognized as a problem in the region, CIAT, in Cali, Colombia began to develop geminivirus resistance in bean for the tropical Americas over a decade ago.[243]

Chino del Tomate and Emerging Geminiviruses of Solanaceae in Mexico

The Chino del tomate (CdTV) geminivirus was the first serious geminivirus pathogen of tomato in Mexico, occurring in west coastal production areas in the 1970's.[244] In contrast to the historically widespread infection of tomato in Old World locations and the lack of a precedent for New World whiteflies to colonize tomato, the occurrence of the CdTV epidemic in Mexico appeared to be an anomaly. The proximity of early planted tomatoes to soybean fields, the source of dispersing whiteflies, appeared to be the major contributing factor, albeit soybean is not a host of the virus.[245] Hence, it was believed that whiteflies acquired CdTV from a weed source en route from soybean to tomato. The disease was a serious limiting factor from until the late 1980's when WFT geminiviruses of tomato and pepper became increasingly important in the area. Interestingly, symptoms of CdTV disease are now rarely observed, and when present, symptoms are apparent in only 1–5% of the plants, whereas the remainder of plants in the field exhibit less severe disease symptoms incited by a complex of poorly characterized geminiviruses including Sinaloa tomato leaf curl virus,[38] serrano golden mosaic virus (SGMV),[246] PHV,[93] and the pepper mild virus from the tigre virus complex,[247] among others.

Why the incidence of disease incited by this virus, once the predominant quasi-species in tomato, is now substantially reduced and yet still present on an annual basis (J. K. Brown, unpublished data), is not known. Interestingly, a virus isolate recently cloned and sequenced from tomato in the same region and tentatively designated tomato leaf crumple (TomLCrV),[248] has an A-component with an IR/CR sequence identical to that of the cloned CdTV-B component (Brown and Bustamante, in preparation), whereas the IR/CR sequence of the TomLCrV-A component varies somewhat from that of its reported homologous B. From these data, it appears that TomLCrV may be a strain of CdTV. The other possibility is that the putative TomLCrV is actually composed of A and B components of two different strains, composed of the A component of CdTV and the B component of an uncharacterized strain, both having been present in the original material. Interestingly, the cloned A and B components of TomLCrV are infectious, thus meeting the requirement for transactivation in one direction.

Vector Biotype-Virus Interactions, and New Diseases

The results from a recent study[249] with the Jatropha biotype and the Jatropha mosaic virus (JMV) indicate that JMV is not transmitted exclusively by the Jatropha biotype, as previously demonstrated.[196] When bean plants were biolistically inoculated[250,251] with JMV to establish a systemic infection, the B biotype (capable of feeding on bean) transmitted JMV with great efficiency from bean to bean. However, the Jatropha biotype could not feed on bean and died after a short exposure to the plant. When the B biotype was allowed free access to JMV-infected Jatropha plants, however, this whitefly survived for several hours and did feed for a short time, as evidenced by the development of JMV symptoms in inoculated plants. These results indicate that the barrier involved in the mobilization of JMV does not reside in limitations of virus-biotype specificity, but in the restricted ability of the Jatropha biotype to exclusively colonize *J. gossypifolia* and its closely related species.[196,249] Interestingly, the JMV was identified as the causal virus in diseased passionvine plantations in the Caribbean Basin,[252] having been mobilized from Jatropha to passionvine plants by either the Jatropha biotype or the B biotype, but ultimately spread throughout the plantation by way of the polyphagous B biotype vector.[249] It is possible that like JMV, other geminiviruses confined to certain host species due to whitefly vector-host plant barriers, also have the potential to be mobilized to other hosts if a suitable whitefly vector is available.

From these and other studies, it is clear that whitefly-host and geminivirus-vector interactions dictate if and when geminiviruses are mobilized from plant to plant, and the rate of disease spread as well. The introduction of the B biotype into the Americas and Caribbean Basin is likely the most important factor in the increase in geminivirus disease epidemics in the region. The altered behavior, broader host range, the ability to disperse over long distances in irrigated desert agroecosystems, the attainment of unprecedented population densities, multiple resistance to insecticides,[253,254] and the ability to transmit New and Old World geminiviruses[78] are among the characteristics that have given this exotic biotype an unprecedented advantage over some of the indigenous *B. tabaci* populations within the region. In Mexico and some sites in Central America and the Caribbean Basin (Belize, Dominican Republic, Guatemala, Honduras), the introduction of the B biotype during the last decade, was preceded by an expansion of acreages devoted to monoculture production of high cash crops for local industries and export purposes. The result has been a steady increase in geminivirus disease incidence linked to increased population densities of *B. tabaci* indigenous to the region.

Emerging Whitefly Biotypes and New Quasi-Species

The specific relationship that has developed between geminiviruses and their whitefly vector appear to have a sustained history of association that has persevered despite the apparent speciation of the whitefly vector. This process is apparently in response to selection forces driven by behavior, host plant associations, environmental conditions, and factors associated with human intervention. What affect, then, does exposure to one or several different biotypes have on geminiviruses that have either extremely broad host ranges, such as the recently described TYLCV-Is strain,[81] or on those that infect a single or only several hosts such as JMV or ACMV, respectively? A case in point is the newly described strain of TYLCV in Israel (TYLCV-IS) which has a much broader host range than previously described, and comparison of the nucleotide sequence with the type strain of TYLCV-Is indicates the two are distinct viruses.[41,81] The new virus is less virulent in tomato than the type TYLCV-Is isolate described from the same area more than thirty years ago,[82] despite the observation that the continual maintenance of high vector densities might have been predicted to give rise to more virulent quasi-species. The implication of this putative shift in virus quasi-species to host plant resistance programs remains to be investigated.

Squash Leaf Curl Virus (SqLCV) Complex: Possible Strain Selection

SqLCV is the only known WFT geminivirus of cucurbits in the New World [4], and has been recognized as a serious economic constraint since its discovery in the Southwestern US in the early 1980's.[43,69,255,256,257] SqLCV wildtype isolates are often composed of a mixture of at least two distinct viral genotypes, SqLCV-E and SqLCV-R. The genomes of these two strains have been cloned and sequenced,[258] and infectious clones have been utilized to study aspects of the molecular biology. Comparison of the DNA sequences of these two strains indicates that the two sequences vary at several sites, for example, in the common regions and in the BR-1 ORFs, among others.[258] The significance of these differences in terms of virus host range phenotype, virulence, or pathogenicity is not well understood, but it is possible to predict from these viral sequences, that competent *trans*-activation of replication is possible between the two strains, and infectivity assays using various combinations of infectious cloned components have corroborated this prediction.[258] Among the noted differences in genomic sequences of the two strains that have been investigated is the 13 base deletion in the common region of SqLCV-R (relative to the E sequence) that is located one nucleotide upstream from the TATA element responsible for transcription of the AL-1 gene. It has been suggested that this mutation reduces rep protein levels due to inefficient transcription of AL-1, leading to reduced replication of the R strain. When the two virus strains occur in mixed infection in certain hosts, the rep protein of the SqLCV-E can enhance or boost the replication of the SqLCV-R.[259] These results predict that under certain circumstances, the more efficiently replicating SqLCV-E will be selected for as the predominant genotype, and in fact, that the R strain might be reduced to low or undetectable levels in the virus pool.

In several preliminary studies, the molecular epidemiological implications of these laboratory-discovered phenomena have been examined under field conditions. Surveys of cucurbit fields conducted in 1986–87[260] using component-specific DNA-DNA hybridization, indicated that the SqLCV-R did not exist in the field as an independent population, but instead, was found in a mixture containing at least one SqLCV-E component. A similar situation was found in follow-up studies using component-specific PCR primers to evaluate the composition of field isolates collected from four different cucurbit genera in the Americas and Mexico from 1978–1994.[79,261] In this study, the E-A and E-B components were found to represent the predominant type, whereas the remaining isolates contained various mixture of three components, and in some cases, all four components. Based on component-specific DNA-DNA hybridization and PCR assays, the R strain was not detected in the absence of the E strain, with two exceptions. The first is the original field isolate of SqLCV from California collected in 1979[255] and in one isolate from squash collected from Nicaragua in 1987 (author, unpublished). Although preliminary, these data suggest that the 'R' strain is not a predominant member of the quasi-species, but it is maintained to some degree in the population In addition, component-specific PCR assays indicated that in several samples, only a single genomic component was present.[261] The latter result suggests that either the component-specific PCR assay does not reliable detect all SqLCV components that are present in samples, or that viral sequences targeted by PCR primers have varied over time. Regardless, it is unclear why the SqLCV-R has survived at all, when the E strain appears to have a competitive edge for replication over the R strain, potentially resulting in an extinction event of the R strain.

One possibility is that the R strain either confers some advantage, or has a neutral effect when present in mixed infection with the E strain. What factors might be responsible for maintaining the R strain at all, when the E strain is known to be capable of replicating more efficiently (at least in bean and squash) are as yet unidentified. Other possible explanations include differential host effects on replication and/or on other virus-host interactions essential for establishing a viable systemic infection, selection by the whitefly vector, or other as yet unidentified helper-type interactions that occur between the two strains of SqLCV. That SqLCV-R strain can rely on the SqLCV-E for replication enhancement suggests an interesting system by which to address epidemiological questions about the genetic composition of this geminivirus strain complex. These findings emphasize the importance of knowledge about geminivirus genotypes and their distribution, and the need to determine the specific factors and forces involved in selection of the predominant strain or quasi-species that comprise virus populations.

Pseudorecombination Between Bean Dwarf Mosaic and Tomato Mottle Viruses: Strains of the Same Quasi-Species?

In the event that transactivation of replication between extant geminiviruses or strains is accepted as a criterion for species and strain designations, several viruses currently considered as separate species will be redefined as strains, thereof. One notable example from the repertoire of geminiviruses examined at this level is the demonstrated pseudorecombination between ToMoV (Florida) BDMV (Colombia),[242] two viruses presently considered to be distinct geminiviruses.[122] Additional evidence may be needed to corroborate the conclusion that these viruses are strains of the same species. For example, both isolates infect bean, do not induce golden mosaic symptoms in bean as do the classical bean golden mosaic virus isolates from the Caribbean Basin,[166] and have highly conserved DNA sequences on both A and B components, collectively indicative of a close relationship. Interestingly, the BDMV of bean in Brazil was reported by Costa (1975) to be the same as the Brazilian AbMV, the causal virus of infectious chlorosis of Malvaceae in the region. DNA hybridization at high stringency using the B components as probes, and comparison of AbMV, BDMV, CdTV and ToMoV at the nucleotide sequence level indicate all four viruses fall within the same subcluster, suggesting an extremely high degree of similarity between these viruses, relative to other geminiviruses (J. K. Brown, unpublished data). Further, CdTV and the recently described TomLCrV[248] (possibly a strain of CdTV?) also fall into this subcluster (Brown, unpublished data). Indeed, these virus isolates may be among the first candidates for consideration as strains of the same virus specie(s). As improved methods give rise to new knowledge and generation of more sophisticated theories on virus relationships and evolution of geminiviruses, several well-characterized geminiviruses presently considered as distinct species will be redefined as strains.

DISEASE CONTROL STRATEGIES

Feasible disease control strategies involve a variety of approaches based on understanding the different components of whitefly-geminivirus-host plant interactions. Other current publications that deal with the specifics of some of the various approaches are emerging, making necessary only brief coverage of these concepts relative to a holistic approach to disease management. Also presented are ideas for several additional approaches that could become reality providing relevant information is gathered and integrated for the application.

Host Plant Resistance

Among the strategies that have received the most attention are those involving disease management using host plant resistance,[262,263,264,265,266] either through plant breeding efforts,[267] or genetic engineering involving pathogen-derived resistance[268] using the coat protein or replicase-related proteins, and defective interfering viruses.[121,269] Partial genetic resistance or tolerance has been found for the New World BGMV,[243] the Old World TYLCV,[85,270] and ToMoV in Florida.[271] Advantages of the existing cultivars or breeding material yet to be introgressed into viable cultivars are that crops can potentially be grown under high virus and vector pressures with reasonable yields. Problems involve low level acceptance of these cultivars which deviate from the preferred varieties for commercial purposes, or from land races grown based on strong cultural preferences. Engineered resistance efforts involve coat protein-mediated resistance against Middle Eastern strains of TYLCV,[270] virus-specific resistance through the expression of antisense viral genes of TGMV,[272,273] virus-specific protection by defective subgenomic viral DNA of ACMV,[274] and dominant-defective resistance strategies.[275]

Vector Thresholds and Integrated Management Approaches

A common method of disease control designed to reduce virus inoculum levels and minimize spread is reduction of vector densities to tolerable (threshold) levels by either chemical insecticides, biological control, or integrated approaches involving several methods. Other combinatorial approaches involve virus resistant cultivars, and cultural/chemical management of weeds and infected plants to minimize virus sources. The reliance on insecticides to reduce whitefly densities is no longer a feasible option in most situations involving WFT geminivirus-incited diseases. There is much interest in implementation of biological control of whiteflies using entomophagous fungi, parasitic wasps, and predators, among other agents under consideration. Yet to be ascertained in order that these control strategies are effective, are specific threshold densities of the whitefly vector that can be tolerated, while reducing virus inoculum levels to the degree that crops can be produced despite virus presence.

Disease Control Through Knowledge of Whitefly Biotype Specificity

It is clear that knowledge of whitefly biotype and the associated virus genotypes/phenotypes will permit predictive approaches that facilitate selective control efforts. For example, a heavy infestation of a non-cassava colonizing *B. tabaci* in tomato will not pose a threat to adjacent cassava fields, nor will the monophagous biotype in cassava affect spread of inoculum in tomato. That the viruses themselves do not cross-infect the two crops is also vital information. These examples, although rare are not useful in situations where geminiviruses are transmitted by a polyphagous whitefly biotype upon which few constraints are placed by crop and weed species in an area. It is important to note however, that unnecessary efforts to reduce population densities that were not above threshold levels, or did not have the potential to colonize a crop or transmit a particular geminivirus because of ignorance have created serious problems in management of virus diseases. An understanding of the virus-vector-host complexes in such cases would probably have delayed or prevented disease problems and the development of insecticide resistance in the insect vector.

Interference with Vector-Mediated Transmission

Although not likely as an immediate solution to the problem of disease spread, strategic interference with some critical phase(s) of the transmission process will be invaluable as a disease control measure. As has been pointed out, there is much to learn about the specificity of geminivirus-whitefly interactions and the underlying mechanisms that can be targeted most effectively to achieve interference approaches to vector-mediated transmission. This shortage of knowledge should not prevent the pursuit of such a strategy, but rather stimulate an interest in light of the potential of such a method to not only to reduce virus transmission in the short term, but also feasibly alter disease epidemiology in the long term by reducing virus virulence and minimizing virus-virus interactions that can result in emergence of new quasi-species. Further, as a component of a multifaceted integrated approach, this strategy offers yet another means for diversification in control methods to minimize the emergence of a more successful quasi-species and/or more efficient virus vector, as often happens when implementing a single approach, thereby fostering selection of a more fit pathogen, or arthropod vector.

OVERVIEW AND CONCLUSIONS

Here, the biological parameters of the WFT geminiviruses have been examined at the phenotypic and genotypic levels, with an emphasis on considerations relevant to interactions between ecological and biological characteristics and molecular evolution of Subgroup III Geminiviruses. The purpose of this writing was to provide an overview of the WFT dicot-infecting Geminiviridae, with an emphasis on the selection pressures imposed by the host plant, the insect vector, and the environmental parameters that affect the molecular epidemiology of WFT geminivirus quasi-species. A better understanding of the biology and molecular epidemiology of these virus pathogens and the application of evolutionary principles to elucidate the selection of characters that drive epidemics, should facilitate identification of the weakest links at which logical interferences can be mounted for disease control.

Through this examination, it is hoped that a more cohesive vantage point can be attained from which to approach the goal of reducing the effects of WFT geminiviruses on crop production. Relevant multifaceted, integrated, and combinatorial approaches should include the development of cultivars that exhibit a pyramiding of inherent host plant and engineered virus pathogen-derived resistance, strategies for interference with vector-mediated geminivirus transmission, cultural management practices for biological control of geminivirus diseases and the whitefly vector, and use of chemical and non-chemical deterrents to interfere with completion of the life cycle of the whitefly vector. This goal will only be realized when virus genetic determinants and those underlying factors that affect the expression of such determinants are taken into account.

ACKNOWLEDGMENTS

The author wishes to thank the following colleagues for helpful discussions during the preparation of this manuscript: Dr. G. K. Banks, Mr. I. D. Bedford, Dr. J. Bird, Dr. D. M. Bisaro, Dr. D. R. Frohlich, Mr. R. J. Gill, Mr. P. Jones, Dr. R. Lastra, Dr. S. G. Lazarowitz, Dr. R. F. Rivera-Bustamante, and Dr. R. C. Rosell. Also for clarification of certain points concerning unpublished or recently published data, appreciation is extended to Dr. R. W. Briddon, Dr. C. M. Fauquet, and Dr. E. P. Rybicki.

DEDICATION

This chapter is dedicated to two mentors, now absent in my life, my mother, Elaine Gilchrist Brown, and my colleague Dr. Victoria Marcarian, agonomist by profession, and citizen of the world. Also to my children, Kevin Brown Chandler and Caitlin Chandler Brown, who have come to understand that plant virologists must frequently travel great distances to be of assistance to their clients.

REFERENCES

1. Muniyappa, V. 1980. Whiteflies. In *Vectors of Plant Pathogens,* K. F. Harris and K. Maramorosch, eds., pp. 39–85. Academic, New York.
2. Brown, J. K. 1990. An update on the whitefly-transmitted geminiviruses in the Americas and the Caribbean basin. *FAO Plant Prot. Bull.* **39:**5–23.
3. Brown, J. K. and J. Bird. 1992. Whitefly-transmitted geminiviruses and associated disorders in the Americas and the Caribbean Basin. *Plant Dis.* **76:**220–225.
4. Brown, J. K. 1994. Current status of *Bemisia tabaci* as a pest and virus vector in world agroecosystems. *FAO Plant Prot. Bull.* **42:**3–32.
5. Duffus, J. E. 1987. Whitefly transmission of plant viruses. In *Current Topics in Vector Research,* K. F. Harris, ed., pp. 73–91. Springer-Verlag, New York.
6. Ewald, P. W. 1983. Host-parasite relations, vectors, and the evolution of disease severity. *Ann. Rev. Ecol. Syst.* **14:**465–485.
7. Eigen, M. 1993. The fifth Paul Ehrlich lecture: virus strains as models of molecular evolution. *Med. Res. Rev.* **13:**385–398.
8. Mayr, E. 1994. Driving forces in evolution. In *The Evolutionary Biology of Viruses,* S. S. Morse, ed., pp. 29–50. Raven, New York.
9. Koblet, H. 1993. Viral evolution and and insects as a possible virologic turning table. *In Vitro Cell. Dev. Biol.* **29A:**274–283.
10. Morse, S. S. 1994. *The evolutionary biology of viruses*. Raven, New York.
11. Bird, J. 1975. *Tropical diseases of legumes*. Academic Press, New York.
12. Bird, J. and K. Maramorosch. 1978. Viruses and virus diseases associated with whiteflies. *Adv. Virus Res.* **22:**55–110.
13. Bock, K. R. 1982. Geminivirus diseases in tropical crops. *Plant Dis.* **66:**266–270.
14. Costa, A. S. 1969. Whiteflies as vectors. In *Viruses, Vectors and Vegetation,* K. Maramorosch, ed., pp. 27–49. Wiley, New York.
15. Costa, A. S. 1976. Whitefly-transmitted plant diseases. *Annu. Rev. Phytopathol.* **14:**429–449.
16. Storey, H. H. 1932. Leaf curl of tobacco in southern Rhodesia. *Rhod. Agric.* **29:**186–192.
17. Varma, P. M. 1963. Transmission of plant viruses by whiteflies. *Nat. Inst. Sci. India.* **24:**11–33.
18. Galvez, G. E. and M. Castano. 1976. Purification of the whitefly-transmitted bean golden mosaic virus. *Turrialba.* **26:**205–207.
19. Goodman, R. M. 1977. Single-stranded genome in a whitefly-transmitted plant virus. *Virology.* **83:**171–179.

20. Hamilton, W. D. O., D. M. Bisaro, R. H. A. Coutts, and K. W. Buck. 1983. Demonstration of the bipartite nature of the genome of a single-stranded DNA plant virus by infection with the cloned DNA components. *Nucl. Acids Res.* **11:**7387–7396.
21. Matyis, J. C., D. M. Silva, A. S. Oliveira, and A. S. Costa. 1975. Purifiacao y morfologia do virus do mosaico durado tomateiro. *Summa Phytopathol.* **1:**267–274.
22. Stanley, J. and M. R. Gay. 1983. Nucleotide sequence of cassava latent virus DNA. *Nature.* **301:**260–262.
23. Abad, J. A. and J. W. Moyer. 1991. Comparison of the capsid protein cistron (CP) from two serologically distinct strains of sweetpotato feathery mottle virus (SPFMV). *Phytopathology.* **81:**1184.
24. Hatta, T. and R. I. B. Francki. 1979. The fine structure of chloris striate mosaic virus. *Virology.* **92:**428–435.
25. Navot, N., E. Pichersky, M. Zeidan, D. Zamir, and H. Czosnek. 1991. Tomato yellow leaf curl virus, a whitefly-transmitted geminivirus with a single genomic component. *Virology.* **185:**151–161.
26. Stanley, J. 1983. Infectivity of the cloned geminivirus genome requires sequences from both DNAs. *Nature.* **305:**643–645.
27. Cohen, S. and R. Ben-Joseph. 1986. Preliminary studies of the distribution of whiteflies *Bemisia tabaci*, using fluorescent dust to mark the insects. *Phytoparasitica.* **14:**152–153.
28. Bisaro, D. M., ed. *Geminivirus replication.* Eukaryotic DNA Replication, ed.^eds. M. DePamphilis. Vol. in press. 1996, Cold Spring Harbor Press**:** New York.
29. Davies, J. W. and J. Stanley. 1989. Geminivirus genes and vectors. *Trends Genet.* **5:**77–81.
30. Fauquet, C. and D. Fargette. 1987. *African cassava mosaic disease and its control.* Editions de l'ORSTOM, Institut Francais de Recherche Scientifique pour le Developpement en Cooperation, Paris.
31. Harrison, B. D. 1985. Advances in geminivirus research. *Annu. Rev. Phtopathol.* **23:**55–82.
32. Lazarowitz, S. G. 1987. The molecular characterization of geminiviruses. *Plant Mol. Biol. Reporter.* **4:**177–192.
33. Lazarowitz, S. G. 1992. Geminiviruses: Genome structure and gene function. *Crit. Rev. Plant Sci.* **11:**327–349.
34. Stanley, J. 1985. The molecular biology of geminiviruses. *Adv. Virus Res.* **30:**139–177.
35. Rybicki, E. P. 1994. A phylogenetic and evolutionary justification for three genera of geminiviridae. *Arch. Virol.* **139:**49–77.
36. Padidam, M., R. N. Beachy, and C. F. Fauquet. 1995. Classification and identification of geminiviruses using sequence comparisons. *J. Gen. Virol.* **76:**249–263.
37. Rybicki, E. P. 1994. ICTV plant virus subcommittee. *Report of the geminivirus study group.* **El Ejido, Almeria, Spain, 1994:**.
38. Brown, J. K., A. M. Idris, and D. C. Fletcher. 1993. Sinaloa tomato leaf curl virus, (Stlcv), a new geminivirus infecting tomato and pepper in northwestern Mexico. *Plant Dis.* **77:**1262.

39. Polston, J. E., E. Hiebert, R. J. McGovern, P. A. Stansly, and D. J. Schuster. 1993. Host range of tomato mottle geminivirus, a new geminivirus infecting tomato in Florida. *Plant Dis.* **77:**1181–1184.
40. McGovern, R. J., J. E. Polston, G. M. Danyluk, E. Hiebert, and A. M. Abouzid. 1994. Identification of a natural weed host of tomato mottle geminivirus in Florida. *Plant Dis.* **78:**1102–1106.
41. Cohen, S. and Y. Antignus. 1993. Tomato yellow leaf curl virus, a whitefly born geminivirus of tomatoes. *Adv. Dis. Vector Res.* **10:**259–288.
42. Cohen, S. and M. J. Berlinger. 1986. Transmission and cultural control of whitefly-borne viruses. *Agric. Ecosys. Environ.* **17:**89–97.
43. Cohen, S., J. E. Duffus, R. C. Larsen, H. Y. Liu, and R. A. Flock. 1983. Purification, serology, and vector relationships of squash leaf curl virus, a whitefly-transmitted geminivirus. *Phytopathology.* **73:**1669–1673.
44. Clark, M. A., L. Baumann, and P. Baumann. 1992. Sequence analysis of an aphid endosymbiont DNA fragment containing {IrpoB} (ß-subunit of RNA Polymerase) and portions of {IrplL} and {IrpoC}. *Current Microbio.* **25:**283–290.
45. Butler, G. D., Jr. and T. J. Henneberry. 1986. *Bemisia tabaci* (Gennadius), a pest of cotton in the southwestern United States. *USDA Agr. Tech. Bull.* **1701:**1–19.
46. Hoefert, L. L. 1987. Association of squash leaf curl virus with nuclei of squash vascular cells. *Phytopathology.* **77:**1596–1600.
47. Jeske, H. and G. Schuchalter-Eicke. 1984. The abutilon mosaic virus (AbMV) in different leaf tissues of several host species of Malvaceae. *Phytopathol. Z.* **109:**353–362.
48. Kim, K. S. and R. Carr. 1982. Characteristic ultrastructure and cytochemistry of plant cells infected with whitefly-transmitted geminiviruses. *Workshop on Plant Virus Detection.* **Agric. Exp. Stn., Rio Piedras, Puerto Rico:**25–33.
49. Osaki, T. and T. Inouye. 1978. Resemblance in morphology and intranuclear appearance of viruses isolated from yellow dwarf diseased tomato and leaf curl diseased tobacco. *Ann. Phytopath. Soc. Japan.* **44:**167–178.
50. Rushing, A. E., G. Sunter, W. E. Gardiner, R. R. Dute, and D. M. Bisaro. 1987. Ultrastructural aspects of tomato golden mosaic virus infection in tobacco. *Phytopathology.* **77:**1231–1236.
51. Tsao, P. W. 1963. Intranuclear inclusion bodies in the leaves of cotton plants infected with leaf crumple virus. *Phytopathology.* **53:**243–244.
52. Kim, K. S. and E. M. Flores. 1979. Nuclear changes associated with euphorbia mosaic virus transmitted by the whitefly. *Phytopathology.* **69:**980–984.
53. Hoefert, L. L. 1983. Ultrastructure of *Cucurbita* spp. infected with whitefly-transmitted squash leaf curl virus. *Phytopathology.* **73:**790.
54. Christie, R. J., N.-J. Ko, J. B. W. Falk, H. Hiebert, R. Lastra, J. Bird, and K. S. Kim. 1986. Light microscopy of geminivirus-induced nuclear inclusion bodies. *Phytopathology.* **76:**124–126.
55. Kim, K. S., T. L. Shock, and R. M. Goodman. 1978. Infection of *Phaseolus vulgaris* by bean golden mosaic virus: Ultrastructural aspects. *Virology.* **89:**22–33.

56. Kim, K. S. and R. W. Fulton. 1984. Ultrastructure of *Datura stramonium* infected with a *Euphorbia* virus suggestive of a whitefly-transmitted geminivirus. *Phytopathology.* **74:**236–241.
57. Horvat, F. and M. Verhoyen. 1981. Cytological modifications and presence of virus-like particles in cells of *Nicotiana benthamiana* (Domin) and *Manihot utilissima* (Pohl) infected with the geminivirus isolated from cassava infected with the cassava African mosaic disease. *Parasitica.* **37:**119–130.
58. Jeske, H. and G. Werz. 1980. Cytochemical characterization of plastidal inclusions in abutilon mosaic-infected *Malva parviflora* mesophyll cells. *Virology.* **106:**155–158.
59. Jeske, H., D. Menzel, and G. Wertz. 1977. Electron microscopic studies on intranuclear virus-like inclusions in mosaic- diseased *Abutilon sellowianum* Reg. *Phytopath. Z.* **91:**1–13.
60. Kim, K. S. and K. W. Lee. 1992. Geminivirus-induced macrotubules and their suggested role in cell-to-cell movement. *Phytopathology.* **64:**664–669.
61. Lastra, R. and F. Gil. 1981. Ultrastructural host cell changes associated with tomato yellow mosaic. *Phytopathology.* **71:**524–528.
62. Russo, M., S. Cohen, and G. P. Martelli. 1980. Virus-like particles in tomato plants affected by the yellow leaf curl disease. *J. Gen. Virol.* **49:**209–213.
63. Saikia, A. K. and V. Muniyappa. 1989. Ultrastructural changes in phloem cells of leaf curl-affected tomato from India. *J. Phytopathol.* **124:**1–6.
64. Thongmeearkom, P., Y. Honda, Y. Saito, and R. Syamananda. 1981. Nuclear ultrastructural changes and aggregates of viruslike particles in mungbean cells affected by mungbean yellow mosaic disease. *Phytopathology.* **71:**41–44.
65. Groning, B. R., A. Abouzid, and H. Jeske. 1987. Single-stranded DNA from abutilon mosaic virus is present in the plastids of infected *Abutilon sellovianum. Proc. Natl. Acad. Sci. USA.* **84:**8996–9000.
66. Horns, T. and H. Jeske. 1991. Localization of abutilon mosaic virus (AbMV) DNA within leaf tissue by *in situ* hybridization. *Virology.* **181:**580–588.
67. Carr, R. J. and K. S. Kim. 1983. Evidence that bean golden mosaic virus invades non-phloem tissue in double infections with tobacco mosaic virus. *J. gen. Virol.* **64:**2489–2492.
68. Dawson, W. O. and M. E. Hilf. 1992. Host range determinants of plant viruses. *Annu. Rev. Plant Physiol.* **43:**527–555.
69. Brown, J. K. and M. R. Nelson. 1986. Whitefly-borne viruses of melons and lettuce in Arizona. *Phytopathology.* **76:**236–239.
70. Brown, J. K. and M. R. Nelson. 1987. Host range and vector relationships of cotton leaf crumple virus. *Plant Dis.* **71:**522–524.
71. Paximadis, M., M. E. C. Rey, and J. K. Brown. 1996. Cloning and sequence comparison of the core coat protein regions of tobacco leaf curl virus isolates from Zimbabwe, Mexico, and the Caribbean Basin. *Cong. Southern African Soc. Plant Pathol.* in press.
72. Backus, E. A. 1988. Sensory systems and behaviors which mediate hemipteran plant feeding: a taxonomic overview. *J. Insect Physiol.* **34:**151–165.

73. Pollard, D. S. 1955. Feeding habits of the cotton whitefly, *Bemisia tabaci* Genn. (Homoptera, Aleyrodidae). *Ann. Appl. Biol.* **43:**664–671.
74. Stenger, D. C., K. R. Davis, and D. M. Bisaro. 1992. Limited replication of tomato golden mosaic virus DNA in explants of nonhost species. *Mol. Plant-Microbe Interact.* **5:**525–527.
75. Blanco Sanchez, N. and I. Bencomo. 1978. Afluencia de la mosca blanca *Bemisia tabaci*, vector del virus del mosaico dorado, en plantaciones de frijol. *Cien. de la Agric.* **2:**39–46.
76. Cohen, S. and V. Melamed-Madjar. 1978. Prevention by soil mulching of the spread of tomato yellow leaf curl virus transmitted by *Bemisia tabaci* (Gennadius) (Hemiptera: Aleyrodidae) in Isreal. *Bull. Ent. Res.* **68:**465–470.
77. Costa, A. S. 1975. Increase in the population density of *Bemisia tabaci*, a threat of widespread virus infection of legume crops in Brazil. In *Tropical Diseases of Legumes,* J. Bird and K. Maramorosch, eds., pp. 27–49. Academic Press, New York.
78. Bedford, I. D., P. J. Markham, J. K. Brown, and R. C. Rosell. 1994. Geminivirus transmission and biological characterization of whitefly *Bemisia tabaci* biotypes from different geographic regions. *Ann. appl. Biol.* **125:**311–325.
79. Brown, J. K., K. Wendt, and S. D. Wyatt. Year. *Genetic variability of squash leaf curl virus isolates by DNA hybridization and component specific polymerase chain reaction.* In *Curcurbitaceae '94*. South Padre Island, TX USA: Gateway Printing. pp. 5–11.
80. Cohen, J. and G. Loebenstein. 1991. Role of a whitefly-transmitted agent in infection of sweet potato by cucumber mosaic virus. *Plant Dis.* **75:**291–292.
81. Antignus, Y. and S. Cohen. 1994. Complete sequence of an infectious clone of a mild isolate of tomato yellow leaf curl virus (TYLCV). *Phytopathology.* **84:**707–712.
82. Cohen, S. and F. E. Nitzany. 1966. Transmission and host range of the tomato yellow leaf curl virus. *Phytopathology.* **56:**1127–1131.
83. Author, u. d., 1995,
84. Dry, I. B., J. E. Rigden, L. R. Krake, P. M. Mullineaux, and M. Ali Rezaian. 1993. Nucleotide sequence and genome organization of tomato leaf curl geminivirus. *J. Gen. Virol.* **74:**147–151.
85. Kheyr-Pour, A., M. Bendahmane, V. Matzeit, G. P. Accotto, S. Crespi, and B. Gronenborn. 1991. Tomato yellow leaf curl virus from Sardinia is a whitefly-transmitted monopartite geminivirus. *Nucl. Acids Res.* **19:**6763–6769.
86. Hamilton, W. D. O., V. E. Stein, R. H. A. Coutts, and K. W. Buck. 1984. Complete nucleotide sequence of the infectious cloned DNA components of tomato golden mosaic virus: potential coding regions and regulatory sequences. *EMBO J.* **3:**2197–2205.
87. Elmer, J. S., L. Brand, G. Sunter, W. E. Gardiner, D. M. Bisaro, and S. G. Rogers. 1988. Genetic analysis of the tomato golden mosaic virus II. The product of the AL1 coding sequence is required for replication. *Nucl. Acids Res.* **16:**7043–7061.

88. Rogers, S. G., D. M. Bisaro, R. B. Horsch, R. T. Fraley, N. L. Hoffmann, L. Brand, J. S. Elmer, and A. M. Lloyd. 1986. Tomato golden mosaic virus A component DNA replicates autonomously in transgenic plants. *Cell.* **45**:593–600.

89. Sunter, G., W. E. Gardiner, A. E. Rushing, S. G. Rogers, and D. M. Bisaro. 1987. Independent encapsidation of tomato golden mosaic virus A component DNA in transgenic plants. *Pl. Mol. Biol.* **8**:477–484.

90. Townsend, R., J. Watts, and J. Stanley. 1986. Synthesis of viral DNA forms in *Nicotiana plumbaginifolia* protoplasts inoculated with cassava latent virus (CLV); evidence for the independent replication of one component of the CLV genome. *Nucl. Acids Res.* **14**:1253–1265.

91. Etessami, P., R. Callis, S. Ellwood, and J. Stanley. 1988. Delimitation of essential genes of cassava latent virus DNA 2. *Nucl. Acids Res.* **16**:4811–4829.

92. Etessami, P., K. Saunders, J. Watts, and J. Stanley. 1991. Mutational analysis of complementary-sense genes of African cassava mosaic virus DNA A. *J. Gen. Virol.* **72**:1005–1012.

93. Cohen, S., J. E. Duffus, and H. Y. Liu. 1989. Acquisition, interference, and retention of cucurbit leaf curl viruses in whiteflies. *Phytopathology.* **79**:109–113.

94. Rybicki, E. P. and F. L. Hughes. 1990. Detection and typing of maize streak virus and other distantly related geminiviruses of grasses by polymerase chain reaction amplification of a conserved viral sequence. *J. Gen. Virol.* **71**:2519–2526.

95. Rochester, D. E., W. Kositratana, and R. N. Beachy. 1990. Systemic movement and symptom production following agroinoculation with a single DNA of tomato yellow leaf curl geminivirus (Thailand). *Virology.* **178**:520–526.

96. Arguello-Astorga, G. R., R. G. Guevara-Gonzalez, L. R. Herrera-Estrella, and R. F. Rivera-Bustamante. 1994. Geminivirus replication origins have a group-specific organization of iterative elements: a model for replication. *Virology.* **203**:90–100.

97. Fenoll, C., J. J. Schwarz, D. M. Black, M. Schneider, and S. H. Howell. 1990. The intergenic region of maize streak virus contains a GC-rich element that activates rightward transcription and binds maize nuclear factors. *Pl. Mol. Biol.* **15**:865–877.

98. Fontes, E. P. B., V. A. Luckow, and L. Hanley-Bowdoin. 1992. A geminivirus replication protein is a sequence-specific DNA binding protein. *The Plant Cell.* **4**:597–608.

99. Revington, G. N., G. Sunter, and D. M. Bisaro. 1989. DNA sequences essential for replication of the B genome component of tomato golden mosaic virus. *The Plant Cell.* **1**:985–992.

100. Thommes, P. A. and K. W. Buck. 1994. Synthesis of tomato golden mosaic virus AL1, AL2, AL3 and AL4 proteins *in vitro*. *J. Gen. Virol.* **75**:1827–1834.

101. Harrison, B. D. and D. J. Robinson. 1988. Molecular variation in vector-borne plant viruses: epidemiological significance. *Phil. Trans. R. Soc. Lond. B.* **321**:447–462.

102. Gardiner, W. E., G. Sunter, L. Brand, J. S. Elmer, S. G. Rogers, and D. M. Bisaro. 1988. Genetic analysis of tomato golden mosaic virus: the coat protein is not required for systemic spread or symptom development. *EMBO J.* **7**:899–904.

103. Sunter, G., M. D. Hartitz, S. G. Hormuzdi, C. L. Brough, and D. M. Bisaro. 1990. Genetic analysis of tomato golden mosaic virus: ORF AL2 is required for coat protein accumulation while ORF AL3 is necessary for efficient DNA replication. *Virology*. **179:**69–77.

104. Sunter, G. and D. M. Bisaro. 1992. Transactivation of geminivirus AR1 and BR1 gene expression by the viral AL2 gene product occurs at the level of transcription. *The Plant Cell*. **4:**1321–1331.

105. Pascal, E., A. A. Sanderfoot, B. Ward, R. Medville, R. Turgeon, and S. G. Lazarowitz. 1994. The geminivirus BR1 movement protein binds single-stranded DNA and localizes to the cell nucleus. *Plant Cell*. **6:**995–1006.

106. Wartig, L., E. Ramasso, A. Kheypour, F. DeKouchkovsky, F. Jouanneau, E. Luisoni, G.-P. Accotto, and B. Gronenborn. 1993. Tomato yellow leaf curl virus (TYLCV) encodes a cytoplasmic non-structural protein VI. *IX International Congress of Virology (abstr. P62–11)*. .

107. Azzam, O., J. Frazer, D. De La Rosa, J. S. Beaver, P. Ahlquist, and D. P. Maxwell. 1994. Whitefly transmission and efficient ssDNA accumulation of bean golden mosaic geminivirus require functional coat protein. *Virology*. **204:**289–296.

108. Briddon, R. W., M. S. Pinner, J. Stanley, and P. G. Markham. 1990. Geminivirus coat protein gene 0nt alters insect specificity. *Virology*. **177:**85–94.

109. Harrison, B. D. and A. F. Murant. 1984. Involvement of virus-coded proteins in transmission of plant viruses by vectors. In *Vectors in Virus Biology,* M. A. Mayo and K. A. Harrap, eds., pp. 1–36. Academic Press, London.

110. Harrison, B. D., V. Muniyappa, M. M. Swanson, I. M. Roberts, and D. J. Robinson. 1991. Recognition and differentiation of seven whitefly-transmitted geminiviruses from India, and their relationships to African cassava mosaic and Thailand mung bean yellow mosaic viruses. *Ann. appl. Biol.* **118:**299–308.

111. Macintosh, S., D. J. Robinson, and B. D. Harrison. 1992. Detection of three whitefly-transmitted geminiviruses occurring in Europe by tests with heterologous monoclonal antibodies. *Ann. appl. Biol.* **121:**297–303.

112. Swanson, M. M., A. Varma, V. Muniyappa, and B. D. Harrison. 1992. Comparative epitope profiles of the particle proteins of whitefly-transmitted geminiviruses from nine crop legumes in India. *Ann. appl. Biol.* **120:**425–433.

113. Swanson, M. M., J. K. Brown, B. T. Poulos, and B. D. Harrison. 1992. Genome affinities and epitope profiles of whitefly-transmitted geminiviruses from the Americas. *Ann. appl. Biol.* **121:**285–296.

114. Swanson, M. M. and B. D. Harrison. 1993. Serological relationships and epitope profiles of isolates of okra leaf curl geminivirus from Africa and the Middle East. *Biochimie*. **75:**707–711.

115. Brown, J. K., S. D. Wyatt, and D. R. Frohlich. 1994. The coat protein gene of whitefly-transmitted geminiviruses, preliminary analysis. *Phytopathology*. **84:**1132.

116. Howarth, A. J., J. Caton, M. Bossert, and R. M. Goodman. 1985. Nucleotide sequence of bean golden mosaic virus and a model for gene regulation in geminiviruses. *Proc. Natl. Acad. Sci. USA*. **82:**3572–3576.

117. Hayes, R. J., C. L. Brough, V. E. Prince, R. H. A. Coutts, and K. W. Buck. 1988. Infection of *Nicotiana benthamiana* with uncut cloned tandem dimers of tomato golden mosaic virus DNA. *J. Gen. Virol.* **69:**209–218.

118. Hanley-Bowdoin, L., J. S. Elmer, and S. G. Rogers. 1990. Expression of functional replication protein from tomato golden mosaic virus in transgenic tobacco plants. *Proc. Natl. Acad. Sci. USA.* **87:**1446–1450.

119. Saunders, K., A. Lucy, and J. Stanley. 1991. DNA forms of the geminivirus African cassava mosaic virus consistent with a rolling circle mechanism of replication. *Nucleic Acids Res.* **19:**2325–2330.

120. Stenger, D. C., G. N. Revington, M. C. Stevenson, and D. M. Bisaro. 1991. Replicational release of geminivirus genomes from tandemly repeated copies: evidence for rolling circle replication. *Phytopathology.* **81:**1184.

121. Frischmuth, T. and J. Stanley. 1991. African cassava mosaic virus DI DNA interferes with the replication of both genomic components. *Virology.* **183:**539–544.

122. Gilbertson, R. L., S. H. Hidayat, E. J. Paplomatas, M. R. Rojas, Y. Hou, and D. P. Maxwell. 1993. Pseudorecombination between infectious cloned DNA components of tomato mottle and bean dwarf mosaic geminivirus. *J. Gen. Virol.* **74:**23–31.

123. Fontes, E. P. B., P. A. Eagle, P. S. Sipe, V. A. Luckow, and L. Hanley-Bowdoin. 1994. Interaction between a geminivirus replication protein and origin DNA is essential for viral replication. *J. Biol. Chem.* **269:**8459–8465.

124. Laufs, J., W. Traut, F. Heyraud, V. Matzeit, S. G. Rogers, J. Schell, and B. Gronenborn. 1995. *In vitro* cleavage and joining at the viral origin of replication by the replication protein Rep of tomato yellow leaf curl virus. *Proc. Natl. Acad. Sci. USA.* **92:**3879–3883.

125. Stenger, D. C., G. N. Revington, M. C. Stevenson, and D. M. Bisaro. 1991. Replicational release of geminivirus genomes from tandemly repeated copies: evidence for rolling-circle replication of a plant viral DNA. *Proc. Natl. Acad. Sci. USA.* **88:**8029–8033.

126. Heyraud, F., V. Matzeit, S. Schaefer, J. Schell, and B. Gronenborn. 1993. The conserved nonanucleotide motif of the geminivirus stem-loop sequence promotes replicational release of virus molecules from redundant copies. *Biochimie.* **75:**605–615.

127. Desbiez, C., C. David, A. Mettouchi, J. Laufs, and B. Gronenborn. 1996. Rep protein of tomato yellow leaf curl geminivirus (TYLCV) has an ATPase activity required for viral DNA replication. *Proc. Natl. Acad. Sci. USA.* **in press:**.

128. Sunter, G., M. D. Hartitz, and D. M. Bisaro. 1993. Tomato golden mosaic virus leftward gene expression: Autoregulation of geminivirus replication protein. *Virology.* **195:**275–280.

129. Sunter, G., M. D. Hartitz, S. G. Hormuzdi, D. C. Stenger, and D. M. Bissaro. 1991. Tomato golden mosaic virus gene function. *Phytopathology.* **81:**1184.

130. Hayes, R. J. and K. W. Buck. 1989. Replication of tomato golden mosaic virus DNA B in transgenic plants expressing open reading frames (ORFs) of DNA A: requirement of ORF AL2 for production of single-stranded DNA. *Nucl. Acids Res.* **17:**10213–10222.

131. Sunter, G., M. Hartitz, and D. M. Bisaro. Year. *Biochemical and genetic analysis of geminivirus proteins controlling replication and transcription.* In *First International Symposium on Geminiviruses.* El Ejido, Almeria, Spain: Universidad de Malaga. pp. 47.

132. Elmer, J. S., G. Sunter, W. E. Gardiner, L. Brand, C. K. Browning, D. M. Bisaro, and S. G. Rogers. 1988. Agrobacterium-mediated inoculation of plants with tomato golden mosaic virus DNAs. *Pl. Mol. Biol.* **10:**225–234.

133. Howarth, A. J. and G. J. Vandemark. 1989. Phylogeny of geminiviruses. *J. gen. Virol.* **70:**2717–2727.

134. Hormuzdi, S. G. 1993. *Genetic studies on tomato golden mosaic virus and beet curly top virus.* Ohio State University, Columbus, Ohio.

135. Rigden, J. E., L. R. Krake, M. A. Rezaian, and I. B. Dry. 1994. ORF C4 of tomato leaf curl geminivirus is a determinant of symptom severity. *Virology.* **204:**847–850.

136. Gröning, B. R., R. J. Hayes, and K. W. Buck. 1994. Simultaneous regulation of tomato golden mosaic virus coat protein and AL1 gene expression: expression of the AL4 gene may contribute to suppression of the AL1 gene. *J. Gen. Virol.* **75:**721–726.

137. Jupin, I., F. de Kouchkovsky, F. Jouanneau, and B. Gronenborn. 1994. Movement of tomato yellow leaf curl geminivirus (TYLCV): Involvement of the protein encoded by ORF C4. *Virology.* **204:**82–90.

138. Ingham, D. J. and S. G. Lazarwitz. 1993. A single missense mutation in the BR1 movement protein alters the host range of squash leaf curl geminivirus. *Virology.* **196:**694–702.

139. Frischmuth, T., S. Roberts, A. von Arnim, and J. Stanley. 1993. Specificity of bipartite geminivirus movement proteins. *Virology.* **196:**666–673.

140. Klinkenberg, F. A., S. Ellwood, and J. Stanley. 1989. Fate of African cassava mosaic virus coat protein deletion mutants after agroinoculation. *J. Gen. Virol.* **70:**1837–1844.

141. von Arnim, A., T. Frischmuth, and J. Stanley. 1993. Detection and possible functions of African cassava mosaic virus DNA B gene products. *Virology.* **192:**264–272.

142. Brough, C. L., R. J. Hayes, A. J. Morgan, R. H. A. Coutts, and K. W. Buck. 1988. Effects of mutagenesis *in vitro* on the ability of cloned tomato golden mosaic virus DNA to infect *Nicotiana benthamiana* plants. *J. Gen. Virol.* **69:**503–514.

143. Noueiry, A. O., W. J. Lucas, and R. L. Gilbertson. 1994. Two proteins of a plant DNA virus coordinate nuclear and plasmodesmata transport. *Plant Cell.* **76:**925–932.

144. Von Arnim, A. and J. Stanley. 1992. Determinants of tomato golden mosaic virus symptom development located on DNA B. *Virology.* **186:**286–293.

145. Pascal, E., P. E. Goodlove, L. C. Wu, and S. G. Lazarowitz. 1993. Transgenic tobacco plants expressing the geminivirus BL1 protein exhibit symptoms of viral disease. *The Plant Cell.* **5:**795–807.

146. Lazarowitz, S. G., L. C. Wu, S. G. Rogers, and J. S. Elmer. 1992. Sequence-specific interaction with the viral AL1 protein identifies a geminivirus DNA replication origin. *The Plant Cell.* **4:**799–809.

147. Ingham, D. J., E. Pascal, and S. G. Lazarowitz. 1995. Both bipartite geminivirus movement proteins define viral host range, but only BL1 determines viral pathogenicity. *Virology*. **207**:191–204.

148. Bisaro, D. M., G. Sunter, G. N. Revington, C. L. Brough, S. G. Hormuzdi, and M. Hartitz. 1990. Molecular genetics of tomato golden mosaic virus replication: Progress toward defining gene functions, transcription units and the origin of DNA replication. In *Viral Genes and Plant Pathogenesis,* T. P. Pirone and J. G. Shaw, eds., pp. 89–105. Springer-Verlag, New York.

149. Rogers, S. G., J. S. Elmer, G. Sunter, W. E. Gardiner, L. Brand, C. K. Browning, and D. M. Bisaro. 1989. Molecular genetics of tomato golden mosaic virus. In *Molecular Biology of Plant-Pathogen Interactions,* B. Staskawicz, P. Ahlquist, andO. Yoder, eds., pp. 199–215. Alan R. Liss, New York.

150. Stanley, J. 1991. The molecular determinants of geminivirus pathogenesis. *Seminars in Virology*. **2**:139–149.

151. Saunders, W. S., C. A. Cooke, and W. C. Earnshaw. 1991. Compartmentalization within the nucleus: discovery of a novel subnuclear region. *J. Cell. Biol.* **115**:919–931.

152. Abouzid, A. M., T. Frischmuth, and H. Jeske. 1988. A putative replicative form of the abutilon mosaic virus (gemini group) in a chromatin-like structure. *Mol Gen Genet.* **212**:252–258.

153. Bellows, T. S., Jr. and K. Arakawa. 1986. Modeling the relationship between transient vector densities and plant disease incidence with special reference to *Bemisia tabaci* (Homoptera: Aleyrodidae) and lettuce infectious virus yellows. *J. Econ. Ent.* **79**:1235–1239.

154. Abouzid, A. M., A. Barth, and H. Jeske. 1988. Immunogold labeling of the Abutilon mosaic virus in ultrathin sections of epoxy resin embedded leaf tissue. *J. Ultrastructure and Molecular Structure Res.* **99**:38–47.

155. Hanley-Bowdoin, L., J. S. Elmer, and S. G. Rogers. 1988. Transient expression of heterologous RNAs using tomato golden mosaic virus. *Nucl. Acids Res.* **16**:10511–10528.

156. Pedersen, T. J. and L. Hanley-Bowdoin. 1994. Molecular characterization of the AL3 protein encoded by a bipartite geminivirus. *Virology*. **202**:1070–1075.

157. Petty, I. T. D., K. W. Buck, and R. H. A. Coutts. 1988. Systemic infection of petunia by mechanical inoculation with tomato golden mosaic virus. *Neth. J. Pl. Path.* **94**:3–7.

158. Sunter, G., W. E. Gardiner, and D. M. Bisaro. 1989. Identification of tomato golden mosaic virus-specific RNAs in infected plants. *Virology*. **170**:243–250.

159. Townsend, R., J. Stanley, S. J. Curson, and M. N. Short. 1985. Major polyadenylated transcipts of cassava latent virus and location of the gene encoding coat protein. *EMBO J.* **4**:33–37.

160. Frischmuth, S., T. Frischmuth, and H. Jeske. 1991. Transcript mapping of abutilon mosaic virus, a geminivirus. *Virology*. **185**:596–604.

161. Mullineaux, P. M., J. E. Rigden, I. B. Dry, L. R. Krake, and M. Ali Rezaian. 1993. Mapping of the polycistronic RNAs of tomato leaf curl geminivirus. *Virology*. **193**:414–423.

162. Kallender, H., I. T. D. Petty, V. E. Stein, M. Panico, I. P. Blench, A. T. Etienne, H. R. Morris, R. H. A. Coutts, and K. W. Buck. 1988. Identification of the coat protein gene of tomato golden mosaic virus. *J. Gen. Virol.* **69:**1351–1357.

163. Gibbs, A. 1987. Molecular evolution of viruses; 'trees', 'clocks' and 'modules'. *J. Cell Sci.* **Suppl. 7:**319–337.

164. Cohen, S., J. Keren, I. Harpaz, and R. Bar-Joseph. 1986. Studies of the epidemiology of a whitefly-borne virus, tomato yellow leaf curl virus, in the Jordan Valley. *Phytoparasitica.* **14:**158.

165. Rosell, R. and J. K. Brown. 1994. Mechanisms of geminivirus acquisition and transmission by the whitefly *Bemisia tabaci* (Genn.). *Phytopathology.* **84:**1131–1132.

166. Faria, J. C., R. L. Gilbertson, S. F. Hanson, F. J. Morales, P. G. Ahlquist, A. L. Loniello, and D. P. Maxwell. 1994. Bean golden mosaic geminivirus type II isolates from the Dominican Republic and Guatemala: nucleotide sequences, infectious pseudorecombinants, and phylogenetic relationships. *Mol. Plant Pathol.* **84:**321–329.

167. Torres-Pacheco, I., J. A. Garzon-Tiznado, L. Herra-Estrella, and R. F. Bustamante-Rivera. 1993. Complete nucleotide sequence of pepper huasteco virus, analysis and comparison with bipartite geminviruses. *J. gen. Virol.* **74:**2225–2231.

168. Faria, J. C., R. L. Gilbertson, F. J. Morales, D. R. Russel, P. G. Ahlquist, S. F. Hanson, and D. P. Maxwell. 1990. Sequence of infectious clones of two mechanically transmissable isolates of bean golden mosaic geminivirus. *Phytopathology.* **80:**983.

169. Arguello-Astorga, G. R., L. R. Herrera-Estrella, and R. F. Rivera-Bustamante. 1994. Experimental and theoretical definition of geminivirus origin of replication. *Plant Mol. Biol.* **26:**553–556.

170. Brown, J. K. 1989. The development of non-radioactive sulfonated probes for the detection and identification of whitefly-transmitted geminiviruses. In Proc. IVth International Plant Virus Epidemiology Workshop. Montpellier, France, 3–8 September 1989: ORSTOM. pp. 276–279.

171. Wyatt, S. D. and J. K. Brown. 1996. Detection of geminiviruses in aqueous extracts by polymerase chain reaction. *Phytopathology.* **in press:**.

172. Gilbertson, R. L., M. R. Rojas, D. R. Russell, and D. P. Maxwell. 1991. Use of the asymmetric polymerase chain reaction and DNA sequencing to determine genetic variability of bean golden mosaic geminivirus in the Dominican Republic. *J. Gen. Virol.* **72:**2843–2848.

173. Brown, J. K., S. D. Wyatt, and G. K. Banks. 1996. The core region of the coat protein gene of subgroup III Geminiviridae is an indicator of geographic origin and may be useful to differentiate strains. *Phytopathol.* **86:**in press.

174. Gennadius, P. 1889. Disease of the tobacco plantations in the Trikonia. The aleurodid of tobacco. *Ellenike Georgia.* **5:**1–3.

175. Quaintance, A. L. and A. C. Baker. 1914. Classification of the Aleyrodidae part II. *USDA Tech. Series.* **27:**95–109.

176. Mohanty, A. K. and A. N. Basu. 1987. Biology of the whitefly vector, *Bemisia tabaci* Genn. on four host plants throughout the year. *J. ent. Res.* **11:**15–18.

177. Mound, L. A. 1963. Host-correlated variation in *Bemisia tabaci* (Gennadius) (Homoptera: Aleyrodidae). *Proc. Royal Entomol. Soc. London.* **(A)38:**171–180.

178. Mound, L. A. and S. H. Halsey. 1978. *Whitefly of the World.* Wiley, New York.

179. Russell, L. M. 1957. Synonyms of *Bemisia tabaci* (Gennadius) (Homoptera: Aleyrodidae). *Bull. Brooklyn Entomol. Soc.* **52:**122–123.

180. Brown, J. K., D. R. Frohlich, and R. Rosell. 1995. The sweetpotato/silverleaf whiteflies: Biotypes of *Bemisia tabaci* Genn., or a species complex? *Ann. Rev. Entomology.* **40:**511–534.

181. Costa, H. S. and J. K. Brown. 1991. Variation in biological characteristics and in esterase patterns among populations of *Bemisia tabaci* (Genn.) and the association of one population with silverleaf symptom development. *Ent. exp. appl.* **61:**211–219.

182. Costa, H. S., J. K. Brown, S. Sivasupramaniam, and J. Bird. 1993. Regional distribution, insecticide resistance, and reciprocal crosses between the A and B biotypes of *Bemisia tabaci. Insect Sci. and Applic.* **14:**255–266.

183. Rosell, R. C., I. D. Bedford, P. G. Markham, D. R. Frohlich, and J. K. Brown. 1994. Morphological variation in *Bemisia* populations. *Phytoparasitica.* **22:**312.

184. Bellows, T. S., Jr., T. M. Perring, R. J. Gill, and D. H. Headrick. 1994. Description of a species of *Bemisia* (Homoptera: Aleyrodidae) infesting North American agriculture. *Ann. Entomol. Soc. Am.* **87:**195–206.

185. Perring, T. M., A. D. Cooper, R. J. Rodriguez, C. A. Farrar, and T. S. J. Bellows. 1993. Identification of a whitefly species by genomic and behavioral studies. *Science.* **259:**74–77.

186. Avidov, Z. 1956. Bionomics of the tobacco whitefly (*Bemisia tabaci* Gennad.) in Israel. *Israel Ktavm.* **7:**25–41.

187. Butler, G. D., Jr., T. J. Henneberry, and F. D. Wilson. 1986. *Bemisia tabaci* (Homoptera: Aleyrodidae) on cotton: adult activity and cultivar oviposition preference. *J. Econ. Ent.* **79:**350–354.

188. Gerling, D., A. R. Horowitz, and J. Baumgaertner. 1986. Autecology of *Bemisia tabaci. Agric. Ecosys. Environ.* **17:**5–19.

189. Butler, G. D. J., T. J. Henneberry, and T. E. Clayton. 1983. *Bemisia tabaci* (Homoptera: Aleyrodidae): development, oviposition, and longevity in relation to temperature. *Ann. Entomol. Soc. Am.* **76:**310–313.

190. Coudriet, D. L., N. Prabhaker, A. N. Kishaba, and D. E. Meyerdirk. 1985. Variation in developmental rate on different hosts and overwintering of the sweetpotato whitefly, *Bemisia tabaci* (Homoptera: Aleyrodidae). *Environ. Ent.* **14:**516–519.

191. Bethke, J. A., T. D. Paine, and G. S. Nuessly. 1991. Comparative biology. morphometrics, and development of two populations of *Bemisia tabaci* (Homoptera: Aleyrodidae) on cotton and poinsettia. *Ann. Entomol. Soc. Am.,.* **84:**407–411.

192. Costa, A. S. and A. M. B. Carvalho. 1969. Comparative studies between *Abutilon* and *Euphorbia* mosaic viruses. *Phytopathol. Z.* **38:**129–152.

193. Cock, M. J. W. 1986. Bemisia tabaci-a literature survey on the cotton whitefly with an annotated bibliography. FAO/CAB, Ascot, UK.

194. Costa, H. S., J. K. Brown, and D. N. Byrne. 1991. Host plant selection by the whitefly, *Bemisia tabaci* Genn. (Homoptera: Aleyrodidae) under greenhouse conditions. *J. Appl. Ent.* **112:**146–152.

195. Costa, H. S., J. K. Brown, and D. N. Byrne. 1991. Life history traits of the whitefly, *Bemisia tabaci* (Homoptera: Aleyrodidae) on six virus-infected or healthy plant species. *Environ. Ent.* **20:**1102–1107.

196. Bird, J. 1957. A whitefly-transmitted mosaic of *Jatropha gossypifolia. Agric. Exp. Stn. Univ. Puerto Rico.* **22:**1–35.

197. Costa, A. S. and L. M. Russell. 1975. Failure of *Bemisia tabaci* to breed on cassava plants in Brazil (Homoptera: Aleyrodidae). *Cienc. Cult. (Sao Paulo).* **27:**388–390.

198. Burban, C., L. D. C. Fishpool, C. Fauquet, D. Fargette, and J.-C. Thouvenel. 1992. Host-associated biotypes within West African populations of the whitefly *Bemisia tabaci* (Genn.), (Hom.,Aleyrodidae). *J. Appl. Ent.* **113:**416–423.

199. Bedford, I. D., R. W. Briddon, P. G. Markham, J. K. Brown, and R. C. Rosell. 1992. A new species of *Bemisia* or biotype of *Bemisia tabaci* (Genn.) as a future pest of European agriculture. Proc. Plant Health and the European Single Market. *BCPC Monograph.* **54:**381–386.

200. Brown, J. K. Year. A critical assessment of biotypes of the sweetpotato whitefly in the Americas and adjacent locales from 1989–1992. In Taller Centro Americano y del Caribe Sobre Moscas Blancas. Turrialba, Costa Rica: CATIE. pp. 1–9.

201. Brown, J. K., S. Coats, I. D. Bedford, P. G. Markham, J. Bird, and D. R. Frohlich. 1995. Characterization and distribution of esterase electromorphs in the whitefly, *Bemisia tabaci* (Genn.) (Homoptera: Aleyrodidae). *Biochem. Genet.* **33:**205–214.

202. Gawel, N. J. and A. C. Bartlett. 1993. Characterization of differences between whiteflies using RAPD-PCR. *Insect Molec. Biol.* **2:**33–38.

203. Frohlich, D. R. and J. K. Brown. 1994. Mitochondrial 16S ribosomal subunit as a molecular marker in Bemisia tabaci and implications for population variability. *Phytoparasitica.* **22:**311.

204. Walker, G. P. and T. M. Perring. 1994. Feeding and oviposition behavior of whiteflies (Homoptera: Aleyrodidae) interpreted from AC electronic feeding monitor waveforms. *Ann. Ent. Soc. Amer.* **87:**363–374.

205. Bird, J., J. Sanchez, R. L. Rodriguez, and P. J. Julia. 1975. Rugaceous viruses in Puerto Rico. In *Tropical Diseases of Legumes,* K. Maramorosch and J. Bird, eds., pp. 3–25. Academic Press, New York.

206. Cohen, S. and I. Harpaz. 1964. Periodic, rather than continual acquisition of a new tomato virus by its vector, the tobacco whitefly *Bemisia tabaci* Gennadius. *Ent. Exp. Appl.* **7:**155–156.

207. Mehta, P., J. A. Wyman, M. K. Nakhla, and D. P. Maxwell. 1994. Transmission of tomato yellow leaf curl geminivirus by *Bemisia tabaci* (Homoptera: Aleyrodidae). *J. Econ. Entomol.* **87:**1291–1297.

208. Polston, J. E., P. Bubrick, and T. P. Perring. 1991. Detection of plant virus coat proteins on whole leaf blots. *Anal. Biochem.* **196:**267–270.

209. Zeidan, M. and H. Czosnek. 1991. Acquisition of tomato yellow leaf curl virus by the whitefly *Bemisia tabaci. J. gen. Virol.* **72:**2607–2614.

210. Polston, J. E., A. Al-Musa, T. M. Perring, and J. A. Dodds. 1990. Association of the nucleic acid of squash leaf curl geminivirus with the whitefly *Bemisia tabaci. Phytopathology.* **80:**850–856.

211. Czosnek, H., R. Ber, N. Navot, Y. Antignus, S. Cohen, and D. Zamir. 1989. Tomato yellow leaf curl virus DNA forms in the viral capside,in infected plants and in the insect vector. *Phytopathol. Z.* **125:**47–54.

212. Navot, N., M. Zeidan, E. Pichersky, D. Zamir, and H. Czosnek. 1992. Use of polymerase chain reaction to amplify tomato yellow leaf curl virus DNA from infected plants and viruliferous whiteflies. *Phytopathology.* **82:**1199–1202.

213. Czosnek, H., R. Ber, N. Navot, D. Zamir, Y. Antignus, and S. Cohen. 1988. Detection of tomato yellow leaf curl virus in lysates of plants and insects by hybridization with a viral DNA probe. *Plant Dis.* **72:**949–951.

214. Rosell, R. C., J. E. Lichty, and J. K. Brown. 1995. Ultrastruture of the mouthparts of adult sweetpotato whitefly, *Bemisia tabaci* Gennadius (Homoptera: Aleyrodidae). *Int. J. Insect Morphol. Embryol.* **42:**297–306.

215. Gildow, F. E. 1987. Virus-membrane interactions involved in circulative transmission of luteoviruses by aphids. In *Current Topics in Vector Research*, pp. 93–120. Springer-Verlag, New York.

216. Gildow, F. E. and S. M. Gray. 1993. The aphid salivary gland basal lamina as a selective barrier associated with vector-specific transmission of barley yellow dwarf luteoviruses. *Phytopathology.* **83:**1293–1302.

217. Navot, N., R. Ber, and H. Czosnek. 1989. Rapid detection of tomato yellow leaf curl virus in squashes of plants and insect vectors. *Phytopathology.* **79:**562–568.

218. Janssen, J. A. M., W. F. Tjallingnii, and J. C. van Lenteren. 1989. Electrical recording and ultrastructure of stylet penetration by the greenhouse whitefly. *Ent. Exp. Appl.* **52:**69–81.

219. Weber, H. 1935. Der bau der imago der Aleurodinen. *Zoologica Heft.* **89:**1–71.

220. Cicero, J. M., E. Hiebert, and S. E. Webb. 1995. The alimentary canal of *Bemisia tabaci* (Genn.) and *Trialeurodes abutilonea* (Hald.) (Aleyrodidae: Sternorhynchi: Homoptera), histology, ultrastructure, and correlations to function. *Zoomorphology.* **115:**31–39.

221. Snodgrass, R. E. 1935. *Principles of Insect Morphology.* McGraw Hill Book Company,

222. R. C. Rosell, p. c., 1994,

223. Rochow, W. F. and I. Muller. 1975. Use of aphids injected with virus-specific antiserum for study of plant viruses that circulate in vectors. *Virology.* **63:**282–286.

224. Gildow, F. E. 1982. Coated-vesicle transport of luteoviruses through salivary glands of *Myzus persicae. Phytopathology.* **72:**1289–1296.

225. Gildow, F. E. 1993. Evidence for receptor-mediated endocytosis regulating luteovirus acquisition by aphids. *Phytopathology.* **83:**270–277.

226. Campbell, B. C. 1993. Congruent evolution between whiteflies (Homoptera: Aleyrodidae) and their bacterial endosymbionts based on respective 18S and 16S rDNAs. *Curr. Micro.* **26:**129–132.

227. Campbell, B. C., J. D. Steffen-Campbell, and R. J. Gill. 1994. Evolutionary origin of whiteflies (Hemiptera: Sternorrhyncha: Aleyrodidae) inferred from 18S rDNA sequences. *Insect Mol. Biol.* **3:**73–88.

228. Hull, R. 1989. Movement of viruses in plants. *Ann. Rev. Phytopathol.* **27:**213–240.

229. Brown, J. K., I. D. Bedford, R. W. Briddon, P. J. Markham, and D. Frohlich. 1993. Relative relationships of whitefly-transmitted geminiviruses by DNA hybridization, and nucleotide sequence alignment. *Intervirol.* .

230. Cohen, S. 1967. The occurrence in the body Bemisia tabaci of a factor apparently related to the phenomenon of "periodic acquisition" of tomato yellow leaf curl virus. *Virology.* **31:**180–183.

231. Cohen, S. and F. E. Nitzany. 1963. Identity of viruses affecting cucurbits in Israel. *Phytopathology.* **53:**193–196.

232. Domingo, E. and J. J. Holland. 1994. Mutation rates and rapid evolution of RNA viruses. In *The Evolutionary Biology of Viruses,* S. S. Morse, ed., pp. 161–184. Raven, New York.

233. Butler, G. D. J., W. D. Hutchison, and M. Broza. 1988. Effect of aldicarb treatments to cotton on *Bemisia tabaci* and *Aphis gossypii* populations in Israel. *SW Entomol.* **13:**87–93.

234. Butler, G. D. and T. J. Henneberry. 1983. Sweetpotato whitefly, *Bemisia tabaci,* 1982 Research Results. *Ariz. Agr. Expt. Sta.* **P-59:**113–119.

235. Fauquet, C. and D. Fargette. 1990. African cassava mosaic virus: etiology, epidemiology, and control. *Plant Dis.* **74:**404–411.

236. Fargette, D., M. Jager, C. F. Fauquet, and L. D. C. Fishpool. 1993. Analysis of temporal disease progress of african cassava mosaic virus. *Ecol. Epidemiol.* **84:**91–98.

237. Muniyappa, V., M. M. Swanson, G. H. Duncan, and B. D. Harrison. 1991. Particle purification, properties and epitope variability of Indian tomato leaf curl geminivirus. *Ann. appl. Biol.* **118:**595–604.

238. Gilbertson, R. L., J. C. Faria, S. F. Hanson, F. J. Morales, P. Ahlquist, D. P. Maxwell, and D. R. Russell. 1991. Cloning of the complete DNA genomes of four bean-infecting geminiviruses and determining their infectivity by electric discharge particle acceleration. *Phytopathology.* **81:**980–985.

239. Goodman, R. M. and J. Bird, *Bean golden mosaic virus* 1978, CMI/AAB Descriptions of Virus Diseases:

240. Gilbertson, R. L., S. H. Hidayat, R. T. Martinez, S. L. Leong, J. C. Faria, F. Morales, and D. P. Maxwell. 1991. Differentiation of bean-infecting geminiviruses by nucleic acid hybridization probes and aspects of bean golden mosaic in Brazil. *Plant Dis.* **75:**336–342.

241. Brown, J. K., M. A. Chapman, and M. R. Nelson. 1989. Bean calico mosaic, a new disease of common bean caused by a whitefly-transmitted geminivirus. *Plant Dis.* **73:**81.

242. Morales, F., A. Niessen, B. Ramirez, and M. Castano. 1990. Isolation and partial characterization of a geminivirus causing bean dwarf mosaic. *Phytopathology*. **80**:96–101.

243. Morales, F. J. and A. I. Niessen. 1988. Comparative responses of selected *Phaseolus vulgaris* germ plasm inoculated artificially and naturally with bean golden mosaic virus. *Plant Dis.* **72**:1020–1023.

244. Brown, J. K. and M. R. Nelson. 1984. Geminate particles associated with cotton leaf crumple disease in Arizona. *Phytopathology*. **74**:987–990.

245. Brown, J. K. and M. R. Nelson. 1988. Transmission, host range, and virus-vector relationships of chino del tomate virus, a whitefly-transmitted geminivirus from Sinaloa, Mexico. *Plant Dis.* **72**:866–869.

246. Brown, J. K. and B. T. Poulos. 1990. Serrano golden mosaic virus: A new whitefly-transmitted geminivirus of pepper and tomato in the U.S. *Plant Dis.* **74**:720.

247. Brown, J. K., O. P. Campodonico, and M. R. Nelson. 1989. A whitefly-transmitted geminivirus from peppers with tigre disease. *Plant Dis.* **73**:610.

248. Paplomatas, E. J., P. D. Grieco, M. R. Rojas, D. P. Maxwell, and R. L. Gilbertson. 1992. Geminivirus complexes associated with tomato and pepper diseases in Mexico. *Phytopathology*. **82**:1070.

249. Brown, J. K. and J. Bird. 1994. Introduction of an exotic whitefly *Bemisia* vector facilitates the spread of Jatropha mosaic virus, a geminivirus previously vectored by the Jatropha biotype. *Phytoparasitica*. **22**:325.

250. Brown, J. K. and R. Ryan. 1991. High velocity microprojectile mediated transmission of whitefly-transmitted geminivirus DNA or purified virions to intact plants. *Phytopathology*. **81**:1217.

251. Brown, J. K. and B. T. Poulos. 1990. Infectious DNA from whitefly-transmitted geminiviruses using particle gun inoculation. In VIIIth International Congress of Virology, August 26–31, 1990. Berlin: pp. 465.

252. Brown, J. K., J. Bird, and D. C. Fletcher. 1993. First report of passiflora leaf mottle caused by a whitefly-transmitted geminivirus in Puerto Rico. *Plant Dis.* **77**:1264.

253. Anthony, N., J. K. Brown, P. G. Markham, and R. H. ffrench-Constant. 1995. Molecular analysis of cyclodiene resistance-associated mutations among populations of the sweetpotato whitefly, *Bemisia tabaci*. *Pest. Biochem. Physiol.* **51**:220–228.

254. Byrne, F. J. and A. L. Devonshire. 1993. Insensitive acetylcholinesterase and esterase polymorphism in susceptible and resistant populations of the tobacco whitefly *Bemisia tabaci* (Genn.). *Pestic. Biochem. Physiol.* **45**:34–42.

255. Flock, R. A. and D. E. Mayhew. 1981. Squash leaf curl, a new disease of cucurbits in California. *Plant Dis.* **65**:75–76.

256. Rosemeyer, M. E., J. K. Brown, and M. R. Nelson. 1986. Five viruses isolated from field-grown buffalo gourd, *Cucurbita foetidissima*, HBK a potential crop for semiarid lands. *Plant Dis.* **70**:405–409.

257. Brown, J. K. and M. R. Nelson. 1989. Characterization of watermelon curly mottle virus, a geminivirus distinct from squash leaf curl virus. *Ann. Appl. Biol.* **115**:243–252.

258. Lazarowitz, S. G. and I. B. Lazdins. 1991. Infectivity and complete nucleotide sequence of the cloned genomic components of a bipartite squash leaf curl geminivirus with a broad host range phenotype. *Virology.* **180:**58–69.
259. Lazarowitz, S. G. 1991. Molecular characterization of two bipartite geminiviruses causing squash leaf curl disease: role of viral replication and movement functions in determining host range. *Virology.* **180:**70–80.
260. Polston, J. E., J. A. Dodds, and T. Perring. 1989. Nucleic acid probes for detection and strain discrimination of cucurbit geminiviruses. *Phytopathology.* **79:**1123–1127.
261. Wendt, K., S. D. Wyatt, and J. K. Brown. 1994. Detection of DNA components of two strains of squash leaf curl virus using polymerase chain reaction. *Phytopathology.* **84:**1104.
262. Baulcombe, D. 1989. Strategies for virus resistance in plants. *Trends Genet.* **5:**56–60.
263. Fraser, R. S. S. 1990. The genetics of resistance to plant viruses. *Annu. Rev. Phytopathol.* **28:**179–200.
264. Hull, R. 1994. Resistance to plant viruses: obtaining genes by non-conventional approaches. *Euphytica.* **75:**195–205.
265. Ponz, F. and G. Bruening. 1986. Mechanisms of resistance to plant viruses. *Ann. Rev. Phytopathol.* **24:**355–381.
266. Wilson, T. M. A. 1993. Strategies to protect crop plants against viruses: pathogen-derived resistance blossoms. *Proc. Nat. Acad. Sci. USA.* **90:**3134–3141.
267. Pilowsky, M. and S. Cohen. 1990. Tolerance to tomato yellow leaf curl virus derived from *Lycopersicon peruvianum. Plant Dis.* **74:**248–250.
268. Sanford, J. C. and S. A. Johnston. 1985. The concept of pathogen-derived resistance-deriving resistance genes from the parasite's own genome. *J. Theor. Biol.* **113:**395–405.
269. Fitchen, J. H. and R. N. Beachy. 1994. Genetically engineered protection against viruses in transgenic plants. *Annu. Rev. Microbiol.* **47:**739–763.
270. Kunik, T., R. Salomon, D. Zamir, N. Navot, M. Zeidan, I. Michelson, Y. Gafni, and H. Czosnek. 1994. Transgenic tomato plants expressing the tomato yellow leaf curl virus capsid protein are resistant to the virus. *Biotechnology.* **12:**500–504.
271. Scott, J. W. and J. H. M. Barten, *Tomato mottle geminivirus,* in *Tomato Yellow Leaf Curl Newslett.*1992, p. 3.
272. Bejarano, E. R. and C. P. Lichtenstein. 1992. Prospects for engineering virus resistance in plants with antisense RNA. *Trends Biotechnol.* **10:**383–388.
273. Day, A. G., E. R. Bejarano, K. W. Buck, M. Burrell, and C. P. Lichtenstein. 1991. Expression of an antisense viral gene in transgenic tobacco confers resistance to the DNA virus tomato golden mosaic virus. *Proc. Natl. Acad. Sci. USA.* **88:**6721–6725.
274. Stanley, J., T. Frischmuth, and S. Ellwood. 1990. Defective viral DNA ameliorates symptoms of geminivirus infection in transgenic plants. *Proc. Natl. Acad. Sci. USA.* **87:**6291–6295.
275. Brown, J. K., S. D. Wyatt, and D. Hazelwood. 1984. Irrigated corn as a source of barley yellow dwarf virus and vector in eastern Washington. *Phytopathology.* **74:**46–49.

276. Costa, A. S. 1969. Conditioning of the plant by one virus necessary for systemic invasion of another. *Phytopathol. Z.* **65:**219–230.
277. Frischmuth, T., G. Zimmat, and H. Jeske. 1990. The nucleotide sequence of Abutilon Mosaic Virus Reveals Prokaryotic as well as Eukaryotic features. *Virology.* **178:**461–468.
278. Morris, B., L. Coates, S. Lowe, K. Richardson, and P. Eddy. 1990. Nucleotide sequence of the infectious cloned DNA components of African cassava mosaic virus (Nigerian strain). *Nucl. Acids Res.* **18:**197–198.
279. Morales, F., A. Niessen, B. Ramirez, and M. Castano. 1989. Isolation and partial characterization of a geminivirus causing bean dwarf mosaic. *Phytopathology.* **79:**910.
280. Brown, J. K. and R. B. Hine. 1984. Geminate particles associated with the leaf curl or 'chino' disease of tomatoes in coastal areas of Western Mexico. *Phytopathology.* **74:**844.
281. Dickson, R. C., M. M. Johnson, and E. F. Laird. 1954. Leaf crumple, a virus disease of cotton. *Phytopathology.* **44:**479–480.
282. Monsoor, S., I. Bedford, M. S. Pinner, J. Stanley, and P. G. Markham. 1993. A whitefly-transmitted geminivirus associated with cotton leaf curl disease in Pakistan. *Pak. J. Bot.* **25:**105–107.
283. Muniyappa, V. and G. K. Veeresh. 1984. Plant virus diseases transmitted by whiteflies in Karnataka. *Proc. Indian Acad. Sci.* **93:**397–406.
284. Inouye, T. and T. Osaki. 1980. The first record in the literature of the possible plant virus disease that appeared in "Manyosshu", a Japanese classic anthology, as far back as the eighth century. *Ann. Phytopath. Soc. Japan.* **46:**449–450.
285. Strandberg, J. O., E. Heibert, G. L. Leibee, and A. Abouzid. 1991. A new geminivirus witha broad host ranbge in the Brassicacaeae (*sic* Cruciferae). *Phytopathology.* **81:**1244.
286. Abouzid, A. M., E. Hiebert, and J. O. Strandberg. 1992. Cloning, identification, and partial sequencing of the genomic components of a geminivirus infecting the Brassicaceae. *Phytopathology.* **82:**1070.
287. Osaki, T., H. Kobatake, and T. Inouye. 1979. Yellow vein mosaic of honeysuckle *Lonicera japonica* Thunb., a disease caused by tobacco leaf curl virus in Japan. *Ann. Phytopath. Soc. Japan.* **45:**62–69.
288. Morinaga, T., M. Ikegami, and K. Miura. 1993. The nucleotide sequence and genome structure of mung bean yellow mosaic geminivirus. *Micro. Immunol.* **37:**471–476.
289. Garzon-Tiznado, J. A., I. Torres-Pacheco, J. T. Ascencio-Ibanez, L. Herrera-Estrella, and R. F. Rivera-Bustamante. 1993. Inoculation of peppers with infectious clones of a new geminivirus by a biolistic procedure. *Phytopathology.* **83:**514–521.
290. Brown, J. K., O. Pozo-Campodonico, and M. R. Nelson. 1989. A whitefly-transmitted geminivirus from peppers with tigre disease. *Plant Dis.* **73:**610.
291. Coutts, R. H. A., R. S. Coffin, E. J. F. Roberts, and W. D. O. Hamilton. 1991. The nucleotide sequence of the infectious cloned DNA components of potato yellow mosaic virus. *J. Gen. Virol.* **72:**1515–1520.

292. Bird, J. 1958. Infectious chlorosis of *Sida carpinifolia* in Puerto Rico. *Agric. Exp. Stn. Univ. Puerto Rico*. **26:**1–23.

293. Iwaki, M., P. Thongmeearkom, Y. Honda, and N. Deema. 1983. Soybean crinkle leaf: a new whitefly-borne disease of soybean. *Plant Dis.* **67:**546–548.

294. Brown, J. K. and M. R. Nelson. 1984. Two whitefly-transmitted viruses of melons in the Southwest. *Phytopathology*. **74:**1136.

295. Stenger, D. C., J. E. Duffus, and B. Villalon. 1990. Biological and genomic properties of a geminivirus isolated from pepper. *Phytopathology*. **80:**704–709.

296. Pruthi, H. S. and C. K. Samuel. 1941. Entomological investigations on the leaf-curl disease of tobacco in northern India. *Indian J. Ag. Sci.* **11:**387–409.

297. Yassin, A. M. and M. A. Nour. 1965. Tomato leaf curl diseases in the Sudan and their relation to tobacco leaf curl. *Ann. appl. Biol.* **56:**207–217.

298. Abouzid, A. M., J. E. Polston, and E. Hiebert. 1992. The nucleotide sequence of tomato mottle virus, a new geminivirus isolated from tomatoes in Florida. *J. Gen. Virol.* **73:**3225–3229.

299. Simone, G. W., J. K. Brown, E. Hiebert, and R. C. Cullen. 1990. Geminiviruses associated with epidemics in Florida tomatoes and pepper. *Phytopathology*. **80:**1063.

300. Jones, P., M. H. A. Satter, and N. Al Kaff. 1988. The incidence of virus disease in watermelon and sweet melon crops in the Peoples Republic of Yemen and its impact on cropping policy. *Aspects of Applied Biology*. **17:**203–207.

CHAPTER 6

Viral Avirulence Genes

J. N. Culver

INTRODUCTION

In 1914, Allard[1] noted that challenging certain tobacco species with sap from a mosaic diseased plant resulted in a "destructive rot" at the point of inoculation. This necrotic response in many respects resembled what Stakman,[2] in 1915, termed the hypersensitive reaction (HR) in particular oat varieties upon inoculation with *Puccinia*. Stakman also correlated this cell death response with host resistance.[2] Thus, these studies provided the first indication that there exists a pathogen defense mechanism in higher plants. In the last 80 years, this resistance phenomena has been the focus of a great deal of research involving its biochemistry and genetics. Much of this work has concentrated on understanding HR specificity. For example, what makes one pathogen elicit an HR whereas another does not? These investigations have led to the identification of numerous pathogen components involved in HR elicitation as well as the recent isolation of several plant disease resistance genes. These findings have substantially advanced our understanding of plant-pathogen interactions and provide a basis for expanding the usefulness of the HR.

The HR is an active defense mechanism that is invoked following pathogen invasion. This response is present in most if not all higher plants, targeting resistance against pathogenic bacteria, fungi, viruses, and nematodes. HR activation results in the rapid necrosis of host cells within the area of pathogen invasion. This localized necrotic response is due in part to the induction of a cascade of host defense responses (phytoalexins, hydrolytic enzymes, callose and lignin precursors, oxidative bursts of H_2O_2, inhibitors of viral replication, etc.) that act to confine the pathogen and prevent disease.[3–5] It is important to note that HR specificity does not appear to involve these plant defense responses. For example, viral elicited HRs induce a wide array of defense proteins such as chitinases that may act against fungal pathogens but have little apparent affect on viral replication. Thus, HR specificity is dependent upon the initial interaction between plant and pathogen.

Control of the HR is often in a gene-for-gene manner involving complementarity between a specific host disease resistance gene(s) and a specific pathogen avirulence gene(s).[6–7] Current models for this gene-for-gene interaction predict that avirulence genes either directly or indirectly produce elicitors that are recognized by specific receptors encoded by plant resistance genes, and that recognition triggers a signal transduction pathway leading to the induction of host defense responses.[8–10] Recent molecular cloning and characterization of several resistance genes has added support to the elicitor-receptor model.[11] Of particular interest, cloned resistance genes involved in the HR against bacterial, fungal, or viral pathogens have common domains that are characteristic of molecules involved in signal transduction.[12–14] These features include a P-loop for ATP/GTP binding and a leucine rich region for protein-protein interactions. Such common features suggest a similar underlying mechanism for HR induction. This information further indicates that specificity for the HR resides within the proposed elicitor-receptor interaction.

VIRAL INDUCED HYPERSENSITIVITY

This chapter focuses primarily on our understanding of how viral genes elicit resistance. Within bacterial and fungal systems, the model of a gene-for-gene interaction has been well established, primarily because of the ability to perform genetic studies demonstrating the presence of dominant pathogen avirulence genes that complement dominant plant resistance genes. The simplicity of the viral genome has precluded the use of a classical genetic approach to identify dominant avirulence factors. However, as early as 1938, Holmes[15] determined necrotic local lesion resistance against tobacco mosaic tobamovirus (TMV) was conferred by a single dominant gene in *Nicotiana glutinosa* L. that he designated *N*. Holmes also noted the presence of similar genes in other systems including pepper and eggplant and the potential use of these genes in controlling disease.[16] To date, plant resistance genes conferring localizing resistance have been identified in numerous virus-host systems.[17–19]

Although viral genome simplicity hinders classical genetic studies, it greatly eases the task of identifying viral encoded components responsible for HR elicitation. Several approaches have been taken. Pseudorecombinant viruses, pairing different genomic segments of closely related multicomponent viruses, have been used to identify symptom determinants.[20–21] However, this approach is limited in resolution and only useful for multicomponent viruses. More recently, the ability to clone entire viral genomes and generate infectious nucleic acid from these clones has provided a major advantage for refined mapping of viral avirulence factors.[22–24] Thus, extending the virus life cycle through a bacterial plasmid has allowed the use of molecular approaches to identify viral factors that elicit resistance. In particular, two techniques have proven most effective in identifying viral factors responsible for HR elicitation. The first is the production of chimeric viruses consisting of genomic segments from different viruses. This method is particularly useful in systems with closely related viruses that have distinct host resistance phenotypes. However, in many systems both resistance inducing and/or noninducing viruses do not exist. To overcome this problem, researchers are beginning to utilize heterologous viral vectors for the expression of specific viral components in attempts to assign avirulence functions. For example, a potato virus X expression vector has been used to identify tomato bushy stunt tombusvirus gene sequences that elicit the HR in *Nicotiana* species.[25–26] Regardless of the technique used, we are just beginning to develop an understanding of the mechanisms by which viral components encode specificity for host recognition and resistance. What follows are several specific examples of identified or suspected viral avirulence genes.

TOBAMOVIRUS COAT PROTEIN/*N'* GENE IN TOBACCO

The *N'* gene, originating from *N. sylvestris* Speg. & Comes directs a HR against most tobamovirus strains. TMV strain U1 does not elicit this resistance response, but instead moves systemically inducing mosaic symptoms and disease in *N'* gene plants.[27] TMV U1 mutants that induce the HR can easily be isolated from a population of the virus or through chemical mutagenesis.[28] Coat protein sequences of these mutants revealed the presence of numerous amino acid substitutions.[28–29] However, no direct evidence existed to eliminate other possible genome mutations from being responsible for HR elicitation. In 1987, Saito et al. substituted the coat protein gene from TMV-OM, which does not induce the HR, into the HR-inducing genome of TMV-L.[30] This chimeric virus did not elicit the

N' gene HR. Thus, the determinant(s) responsible for HR elicitation resided within coat protein gene sequences.

Subsequently, Knorr and Dawson[31] cloned and sequenced portions of an HR-eliciting mutant of TMV-U1. Chimeric constructs between mutant and wild-type virus identified a single nucleotide difference (nucleotide 6157, cytosine to uracil) in the coat protein open reading frame (ORF) that was responsible for conferring the ability to elicit the HR. This nucleotide alteration led to an amino acid substitution of Ser to Phe at position 148 in the coat protein. By coincidence, this amino acid substitution was one of several previously described in N' gene eliciting TMV mutants,[28] suggesting that other mutations within the coat protein might also lead to HR elicitation. To address this question, four previously described amino acid substitutions were recreated in the coat protein ORF.[32] These substitutions were the only alterations in the viral genome. Infectious transcripts from these four mutants induced the HR in *N. sylvestris,* demonstrating the ability of multiple changes throughout the coat protein ORF to elicit the HR.

Interestingly, each mutant gave a distinctly different pattern of lesion development.[32] Based on the developmental pattern of necrotic lesions in N' gene plants, three phenotypic groups were identified. "Strong elicitors" induced small lesions, 1–2 mm in diameter, within 2–3 days postinoculation that did not increase in size. "Moderate elicitors" induced lesions, 1–2 mm in diameter, at 4–5 days postinoculation that increased in size, up to 5–10 mm in diameter. "Weak elicitors" induced large necrotic lesions, 4–7 mm in diameter, that appeared 5–7 days postinoculation and continually expanded, eventually collapsing large portions of the inoculated leaves. The active determinant required for *N'* gene recognition could have resided in either the RNA or the protein. However, alteration of the coat protein translational start codon (AUG to AGA) of one HR-inducing mutant resulted in a virus that replicated efficiently, did not produce coat protein, and did not elicit the HR.[33] Therefore, production of altered coat protein during viral replication was required for induction of the HR.

To determine if viral components or replication processes other than coat protein were involved in the induction of the *N'* gene HR, transgenic *N. sylvestris* plants expressing the coat protein ORF of either HR-inducing mutants or the wild-type virus that did not induce the HR were produced.[34] Each coat protein ORF was expressed behind the 35S promoter of cauliflower mosaic virus (CaMV). Transgenic *N. sylvestris* expressing the wild-type coat protein developed normally when compared to nontransgenic control plants. In contrast, transgenic *N. sylvestris* plants expressing elicitor coat proteins displayed varying degrees of growth reduction and necrosis on mature leaves. Similarly transformed callus tissue from *N'* like Samsun *EN* tobacco also displayed an HR phenotype.[35] Thus, an HR-eliciting coat protein alone is sufficient to elicit the HR.

Importance of Coat Protein Structure in HR Induction

The ability of different coat protein amino acid substitutions to elicit the N' gene HR suggests that these substitutions have a similar effect on the protein. It is unlikely that wild-type coat protein acts to suppress the HR and that HR-inducing substitutions interfere with this suppressing ability, because mutants that fail to produce coat protein do not elicit the HR, and transgenic plants that express wild-type coat protein still display the characteristic HR when inoculated with HR-inducing strains of TMV.[34] Additionally, these mutant coat proteins elicit an HR only in *N'* genotype plants.[32] Thus, mutations that lead to the HR must result in a protein configuration that can be recognized by *N'* gene plants.

Fortunately the three-dimensional structure of the TMV coat protein has been resolved to 2.8Å for a 20S disk aggregate, consisting of two stacked protein rings each containing 17 coat protein subunits[36] and at 2.9Å resolution for the TMV virion.[37] The core (middle radius) of the molecule is composed of four right-handed antiparallel alpha-helices, designated left and right slewed and left and right radial. The inner radius of the molecule consists of a short loop connecting the LS and RS helices and a longer loop connecting the LR and RR helices. At the extreme outer radius are the carboxy and amino termini. In the absence of viral RNA, TMV coat protein forms a variety of aggregates depending upon solution conditions.[38] Under cellular conditions the 20S disk aggregate or a short helix containing about 38 subunits predominates.[39–40] Coat protein aggregates are highly ordered with extensive intermolecular interactions both laterally (side-to-side) and axially (up and down). Thus, aggregate forms of coat protein must be considered when examining any structural alteration that elicits the *N'* gene HR.

Involvement of Coat Protein Quaternary Structure

Initially, five independent amino acid substitutions were identified within the coat protein (Val 11 to Met, Pro 20 to Leu, Asn 25 to Ser, Arg 46 to Gly, and Ser 148 to Phe) that led to elicitation of the HR.[31–32] Each of these substitutions was different with no obvious similarities between the type of amino acid exchanged and location within a single coat protein molecule. However, based on computer generated models of coat protein aggregates,[41] these substitutions share a common feature in that all are located within surface interface regions that occur between coat protein subunits. Residues 11, 20, 25, and 46 are all important in lateral interactions, while residue 148 is involved in an axial interaction between disk layers. Thus, alterations within coat protein interface regions appeared to affect elicitation of the *N'* gene HR.

To define further the importance of interface regions a number of additional amino acid substitutions were selectively created throughout the coat protein molecule.[42] Of 19 additional substitutions, 10 resulted in induction of the HR. All of these HR-eliciting amino acid substitutions were located within interface regions between adjacent coat protein subunits (Fig. 6–1). Based on the known coat

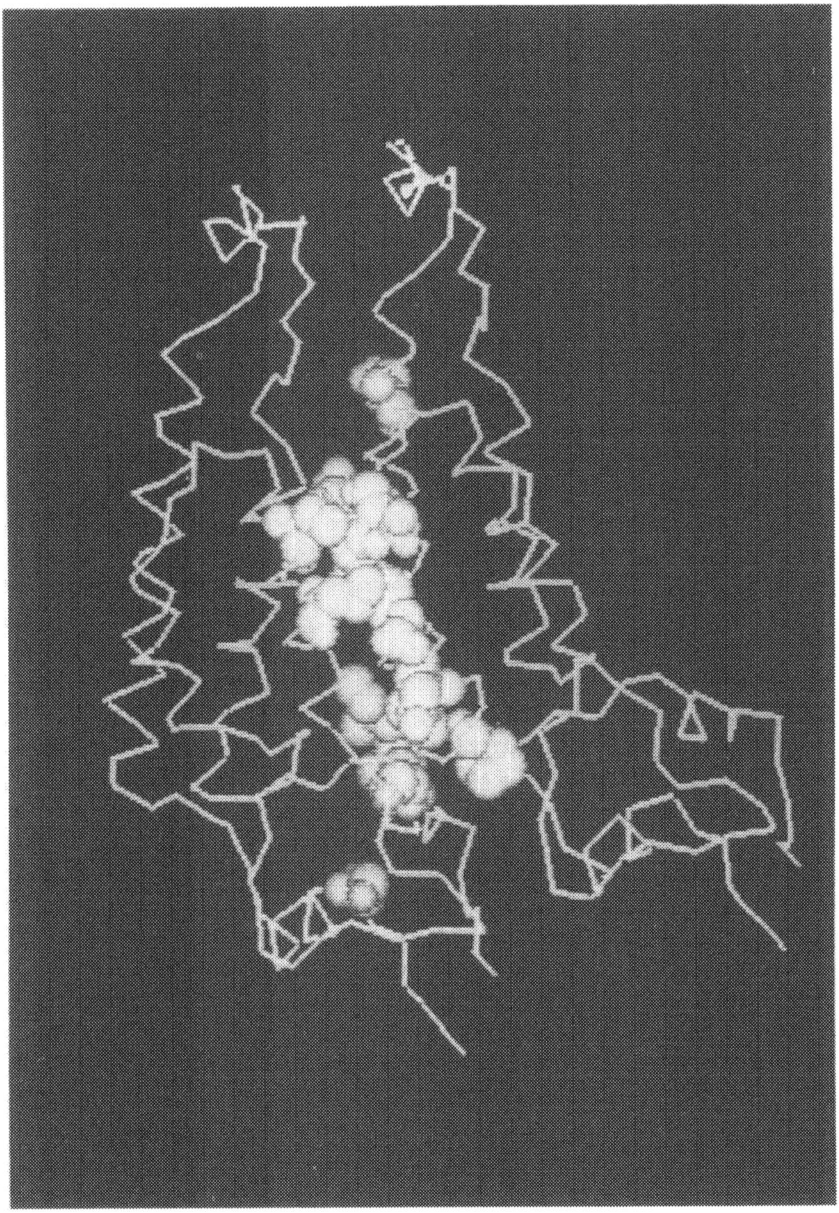

Figure 6–1. Alpha carbon backbone of two tobacco mosaic virus coat protein subunits positioned in aggregate form. Spatial locations of substituted wild-type residues that elicited the *N'* gene hypersensitive response are shown in space-filling form.

protein structure, these substitutions would disrupt either lateral or axial contacts between adjacent molecules. The nine amino acid substitutions that did not elicit the HR were either conservative in change or located outside of interface regions. Thus, amino acid substitutions that predictably interfere with subunit-to-subunit interactions resulted in elicitation of the *N'* gene HR.

To investigate the direct affects of HR-inducing mutations on coat protein quaternary structure the ability of these proteins to form ordered aggregates was examined. Aggregation properties of TMV coat protein in the absence of viral RNA have been well characterized under a variety of conditions.[38] For the wild-type coat protein, ordered aggregates form below pH 9.0 in 100mM Tris buffer with disk-shaped aggregates predominating as the pH is lowered to around 7.0. Electron microscopy of mutant coat proteins, equilibrated to pH 7.0, showed that all were capable of forming these aggregates regardless of their ability to elicit the HR. However, when the pH of these samples was raised to 8.0, close to the limiting conditions for disk stability, only the wild-type and other non-HR-eliciting coat proteins maintained disk structures.

The ability of mutant coat proteins to aggregate into helical structures was also determined. Even in the absence of viral RNA, wild-type TMV coat protein will rapidly form long helices (rods) that remain soluble at pH 5.0.[38] Coat proteins that are deficient in the ability to form rods become insoluble at pH 5.0 and pellet upon centrifugation.[43] Stability of mutant coat protein helical aggregates was determined both by measuring the amount of protein that remained in solution at pH 5.0 and by electron microscopy. Wild-type and non-HR eliciting coat proteins that strongly elicited the HR were not capable of forming helical rods at pH 5.0, while the wild-type and non-HR eliciting coat proteins formed stable helices. A weak elicitor coat protein was also capable of forming helical rods but with slightly reduced levels of soluble protein. In contrast, a moderate elicitor coat protein formed primarily short helical rods with levels of soluble protein about half that of non-HR eliciting coat proteins, while strong elicitor coat proteins formed no soluble helical rods at pH 5.0. These data demonstrate that HR-eliciting coat proteins have reduced quaternary structure stability and that reduction in stability correlates with the elicitor strength of the protein.

The effects of restoring quaternary structure stability to elicitor coat proteins was also examined. This was done by introducing a second amino acid substitution into an existing elicitor coat protein. Each second mutation was designed to either restore the structure disrupted by the elicitor substitution or increase overall stability of coat protein aggregates. For example, changing Arg 46 to Gly resulted in a disruption of an axial interface and moderate elicitation of the HR. A second substitution, Thr 42 to Arg was created to restore the charge and side-chain packing lost by the original Arg 46 to Gly substitution. This double mutation only weakly elicited the HR, indicating that partial restoration of the normal interface structure could reduce the HR. A second set of mutations was designed to increase the overall stability of elicitor coat proteins by the removal of repulsive negatively charged carboxyl-carboxylate pairs that occur between adjacent subunits. Normally, destabilization of quaternary structure introduced by these pairs is an important step in virion disassembly.[37] To strengthen overall quaternary structure members of one carboxyl-carboxylate pair, Glu 50 and Asp 77, were substituted with corresponding noncharged but similar residues Gln and Asn. These substitutions by themselves do not elicit the HR and have been shown to increase the stability of coat protein quaternary structure.[44] However, placement of these substitutions into a strong elicitor coat protein resulted in a reduction in elicitor strength. This indicates that overall strengthening of elicitor coat protein quaternary structure leads to reduced elicitor strength. These findings further support the importance of coat protein quaternary structure in *N'* gene HR elicitation.

Importance of Coat Protein Tertiary Structure

The induction of the *N'* gene HR is not just associated with the reduced ability of coat protein to aggregate. Coat protein must be able to properly fold, at least in part, into its native three-dimensional tertiary structure. This requirement is demonstrated by the inability of elicitor coat protein mutants having internal deletions to elicit the HR,[45] suggesting that certain tertiary structural features of the coat protein must be maintained in order for host recognition to occur. To investigate specific structural features within coat protein that are important for host recognition, second-site amino acid substitutions have been added to an existing strong elicitor coat protein.[46] Second-site substitutions were designed to disrupt coat protein structural features in specific areas of the protein. When created alone in the wild-type nonelicitor coat protein these second-site substitutions do not elicit the HR.

From over 30 second-site substitutions, 14 were found to result in either a complete or reduced loss in coat protein ability to elicit the HR.[46] Several of these mutations were dominant for the loss of HR only at the increased temperature of 29°C, normally permissive for *N'* gene elicitation. Elevated temperature may enhance the ability of some mutations to destabilize structural motifs important for HR elicitation. Interestingly, all of these dominant-negative substitutions lie in and around the right side of the alpha-helical bundle (Fig. 6–2). Substitutions that were not dominant for the loss of HR occurred outside this structural region. Thus, the structural configuration of this area appears to be important for host recognition and could possibly delineate a putative receptor binding site. This region is also buried within the normal wild-type quaternary configuration of coat protein. Additionally, this structural region contains a cluster of residues conserved among the tobamoviruses that elicit the *N'* gene HR[47], presumably because of their importance in maintaining coat protein function and structure. Thus, in order to

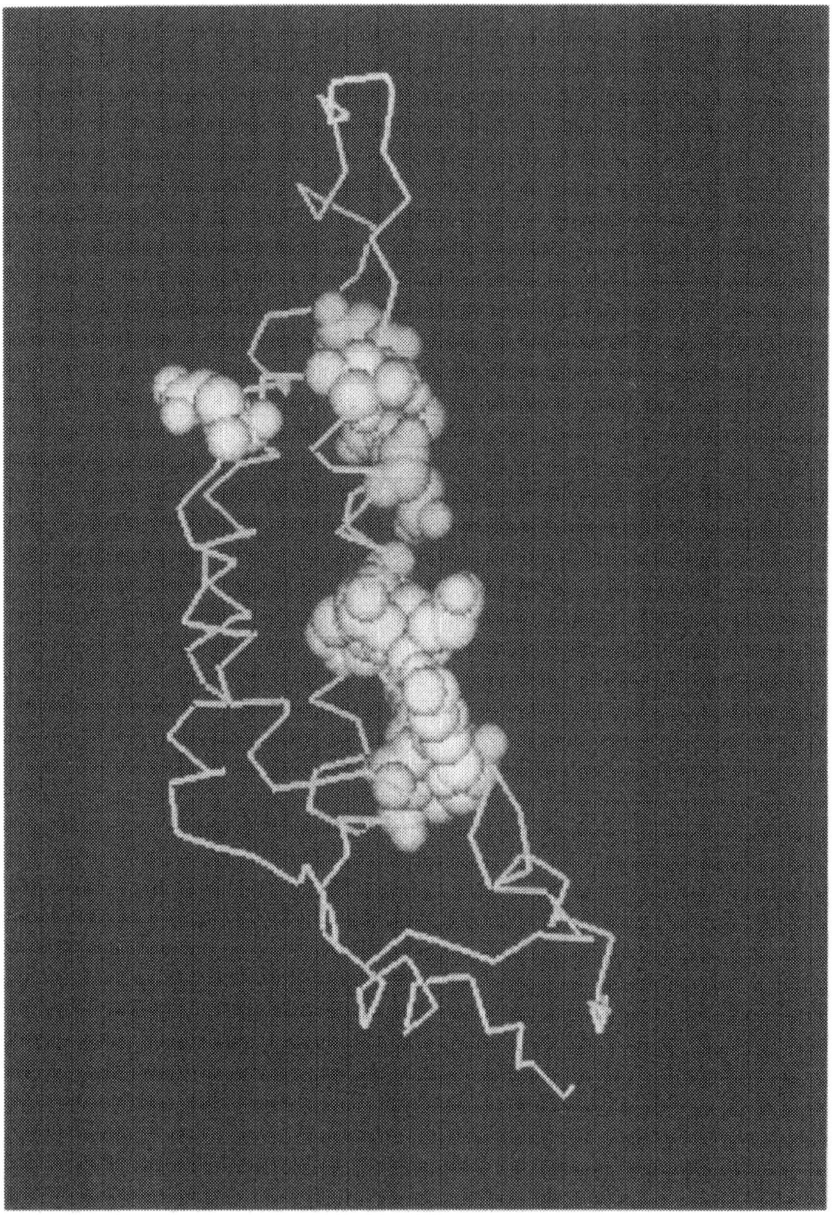

Figure 6–2. Alpha carbon backbone of a single tobacco mosaic virus coat protein subunit. Spatial locations of substituted second-site mutations that result in the loss of elicitor activity are shown in space-filling form.

overcome *N'* gene resistance and still maintain a functional coat protein, TMV-U1 may have evolved a "tighter" quaternary structure to mask this site.

A Mechanism for Coat Protein Elicitation of the *N'* Gene HR

Current evidence indicates that *N'* gene recognition requires structurally intact coat proteins with reduced subunit-to-subunit interactions. However, no correlation exists between a total loss of aggregation and the induction of the HR. Many HR-inducing mutants produce stable virions and protein aggregates under physiological conditions. It appears that mutations that lead to the induction of the HR need only create small changes in the aggregating ability of coat protein. These changes may shift the equilibrium of coat protein aggregation, allowing the accumulation of a threshold level of coat protein form (monomer, dimer, trimer, etc.) that is recognized by the host. Based on this evidence, the following hypothesis for N' gene recognition is proposed: Amino acid substitutions that lead to HR elicitation disrupt the ordered quaternary structure of coat protein and expose a structural motif "receptor binding site" that is normally not accessible for recognition by the plant (Fig. 6–3). This structural motif consists of noncontiguous stretches of amino acids whose relative position within the tertiary structure of coat protein must be maintained. The strength of the HR reaction, strong vs. weak, depends upon the accessibility and structural integrity of the receptor binding site.

TOBAMOVIRUS COAT PROTEIN/L^3 GENE IN PEPPER

The dominant gene, L^3, in *Capsicum* L. (pepper) confers a HR against tobamovirus infection[48]. Berzal-Herranz et al. have recently created chimeric constructs among tobamoviruses with differing abilities to induce these HRs[49]. These studies indicate that the coat protein gene is responsible for eliciting the HR in L^3 peppers. Specifically, a single nucleotide conversion resulting in a coat protein amino acid substitution of Asn to Met at position 138 is sufficient to induce the HR in L^3 peppers. Such an amino acid substitution, leading to HR induction, suggests a similar mechanism of induction for L3 as in the *N'* gene of tobacco. It will be interesting to analyze the structural consequences of L^3 eliciting coat protein substitutions and compare these to the results obtained from *N'* gene eliciting substitutions.

TOBAMOVIRUS REPLICASE/*N* GENE IN TOBACCO

The N gene from *Nicotiana glutinosa* L. directs an HR against almost all strains of tobamoviruses. This resistance has proven to be one of the most resilient and

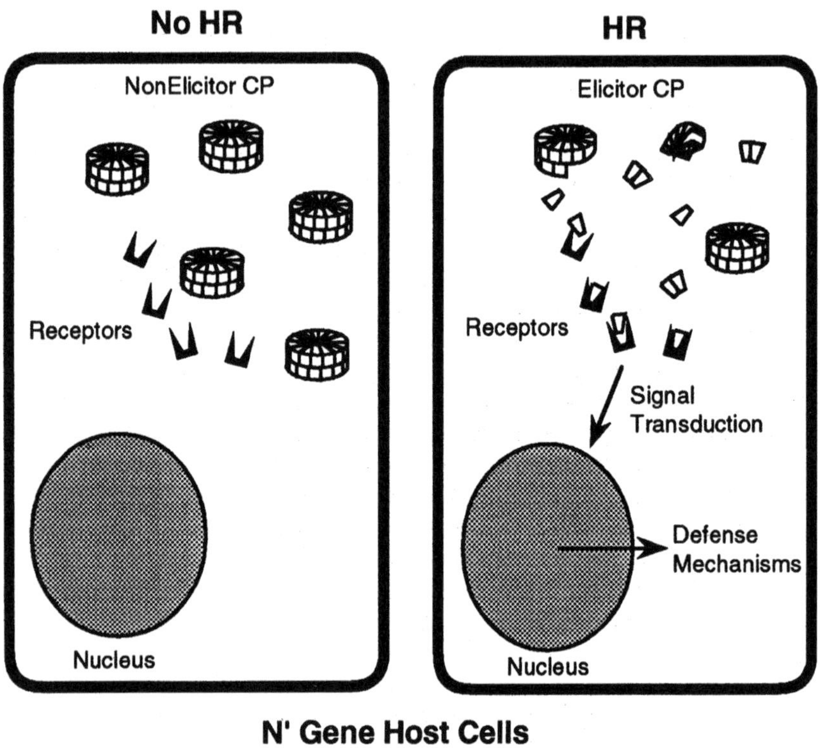

Figure 6–3. Model explaining the effects of elicitor and nonelicitor TMV coat proteins on the induction of the *N'* gene hypersensitive response.

useful responses, both in disease control and as an experimental tool for virus characterization[15]. Coat protein is not involved in eliciting the *N* gene HR, as demonstrated by the ability of TMV mutants missing the entire coat protein ORF to elicit necrotic lesions that are indistinguishable from those induced by the wild-type virus.[45,50] The only strain capable of overcoming *N* gene resistance is TMV-Ob.[51] Padgett and Beachy[52] have recently cloned and sequenced TMV-Ob in an attempt to identify an *N* gene eliciting viral component. In this work, the wild-type TMV 30 kDa movement protein was replaced with the movement protein from TMV-Ob. This hybrid virus still elicited the HR normally. It was also noted that transgenic *N* gene plants expressing movement protein do not display an HR phenotype.[53] Thus, the TMV movement protein, like the coat protein, is not responsible for *N* gene elicitation.

Involvement of TMV Replicase

In further studies, Padgett and Beachy used chemical mutagenesis to obtain a TMV-Ob mutant (ObNL-1) that elicited the *N* gene HR. The HR elicited by ObNL-1 was distinct in both timing and appearance from that elicited by the wild-type virus. Nevertheless this response was specific to hosts carrying the *N* gene. Sequence analysis revealed a single cytosine to uridine transition at nucleotide 3334. This mutation resulted in an amino acid substitution of proline to leucine at position 1089 in the 126 kDa viral replicase protein. Thus, a single mutation within the TMV-Ob genome altered the *N* gene interaction from susceptible to resistance. How this single alteration results in the HR is unclear, but it is possible that this mutation leads to a protein configuration that is recognized by the *N* gene product. Padgett and Beachy speculate that the proline substitution at this position could affect the alpha-helical content of the protein, thus altering its structural configuration. Another possibility is that this mutation effects the production of a replicase product or complex that is then recognized by the *N* gene product. It is also interesting to note that TMV-Ob does induce the *N* gene HR at lower temperatures, 18–20°C.[51] This indicates that the TMV-Ob strain does contain or express the appropriate features for recognition but may, as in the TMV coat protein/*N'* interaction, have evolved a mechanism to mask those features. It will be interesting to determine if other changes in the TMV-Ob 126 kDa protein, differing by 33% from the *N* gene eliciting U1 strain, have an effect on this interaction with regard to the HR-phenotype and temperature sensitivity. This information along with a more detailed understanding of the TMV replication process should hclp elucidate the *N* gene HR elicitor. Additionally, it may be possible to utilize the recently cloned *N* gene product[14] as a probe for isolating the elicitor.

The *N* Gene

The recent cloning of the *N* gene from *N. glutinosa* provides our first look at a host component involved in a viral elicited HR.[14] As indicated earlier, cloned resistance genes, including the N gene, have common domains characteristic of molecules involved in signal transduction. Most notably, the *N* gene shares 49% similarity and 24% identity with the RPS2 gene of *Arabidopsis thaliana,* conferring resistance against the bacterial pathogen *Pseudomonas* syringae.[12–14] This is a remarkable finding, considering that *N* and RPS2 both originate from different plant families and confer HRs against different pathogens. Southern blot analysis also indicates the presence of *N* gene homologues in *Nicotiana* species that do not display a similar HR phenotype.[14] This suggests that a family of structurally similar but functionally different genes exists. Perhaps each homologue directs the HR against a different pathogen elicitor. Another point of interest is that sequence data indicates the *N* gene product to be cytoplasmic, suggesting that virus recognition occurs in the cytosol. A requirement for cytoplasmic recognition is also observed in the *N'*/TMV coat protein interaction in which surface infiltration of elicitor coat protein fails to induce the HR.[34] Cytoplasmic recognition of viral pathogens is not surprising, since viral replication and elicitor production would occur from within the cell. Whether viral elicitors interact directly with the product of the resistance gene or through some intermediate remains to be seen, but certainly the tools to elucidate this interaction from both pathogen and plant sides now exists.

TOBAMOVIRUS MOVEMENT PROTEIN/*TM-2* AND *TM-2*^{2} IN TOMATO

Two genes, *Tm-2* and *Tm-2*^{2}, in the genus *Lycopersicon* have been identified as providing resistance against TMV.[54,55] Interestingly, the responses conferred by these genes can be accompanied by either localized necrosis, resulting in resistance, or systemic necrosis, resulting in severe disease. The type of necrosis is dependent upon the strain of infecting virus, the genotype of the plant, and the environmental conditions, especially temperature. *Tm-2* and *Tm2*^{2} are thought to be allelic;[54,55] however, *Tm-2* resistance is easily overcome by naturally occurring strains of tobamoviruses. Strains that overcome *Tm-2*^{2} resistance lead to systemic necrosis and severe disease, a phenotype that can easily be rogued out under conditions of commercial production. This systemic necrosis suggests that tobamovirus strains that overcome *Tm-2*^{2} do so only partially, perhaps similar to weak elicitor proteins in the TMV coat protein/*N'* system.

Several lines of evidence suggest the involvement of the viral 30kDa movement protein in overcoming *Tm-2* and *Tm-2*^{2} resistance. Both genes provide protection only in plants and not at the level of protoplasts.[56] Protection is also observed as a restriction in the movement of TMV to initially infected cells in both *Tm-2* and *Tm-2*^{2} plants.[57]

Meshi et al. has examined the nucleotide sequence of two independent TMV mutants, Ltb1 and C32, capable of overcoming *Tm-2* resistance.[58] Sequence comparisons between resistance-breaking mutants and the wild-type virus revealed several amino acid differences within the virus 30kDa protein. For mutant Ltb1, substitutions of Cys 68 to Phe and Glu 133 to Lys were required to overcome *Tm-2* resistance, while in mutant C32 substitutions of Glu 52 to Lys and Glu 133 to Lys were required. These data demonstrate involvement of the TMV movement protein in overcoming *Tm-2* resistance.

In similar experiments, Weber et al. have compared nucleotide sequences between a *Tm-2*^{2} resistance-breaking strain, tomato mosaic tobamovirus-2^2 (ToMV), and the wild-type ToMV.[59] Four amino acid substitutions were initially identified, Lys 130 to Glu, Gly 184 to Glu, Ser 238 to Arg, and Lys 244 to Glu. The introduction of various combinations of these mutations into wild-type ToMV revealed that the Arg 238 and Glu 244 mutations were responsible for overcoming *Tm-2*^{2} resistance. These mutations occur within the highly variable carboxy-terminal domain of the 30kDa protein, a region that is dispensable for the cell-to-cell movement function of the protein.[60] Based on these observations Weber et al. predict the presence of a specific interaction between the *Tm-2*^{2} gene and the tobamovirus 30–kDa protein, leading to the activation of defense mechanisms. Recently, it was reported that the expression of the wild-type ToMV 30kDa protein in transgenic *Tm-2*^{2} tomato results in a necrotic reaction.[61] This evidence demonstrates that the 30kDa movement protein acts as an elicitor molecule for induction of the HR in *Tm-2*^{2} tomatoes.

POTATO POTEXVIRUS X COAT PROTEIN/*RX* AND *NX* IN POTATO

The coat protein gene of potato potexvirus X (PVX), like that of TMV, has been identified as an avirulence factor involved in several HRs. Baulcombe and colleagues have begun to define these interactions and have revealed an interesting diversity in the ability of this viral component to elicit plant resistance.

Nx Derived Resistance

The *Nx* resistance gene in *Solanum* species confers a HR against group 1 and group 3 strains of potato virus X (PVX).[62] Development of induced necrosis can vary from local to systemic depending upon viral inoculum concentrations and environmental conditions[62]. Through coat protein replacement experiments, Kavanagh et al. have mapped the *Nx* resistance-breaking ability of specific PVX strains to the coat protein ORF.[64] Santa Cruz and Baulcombe[65] further identified a single nucleotide change leading to an amino acid substitution, Glu78 to Pro, in the coat protein of the resistance-breaking strain PVXDX4 as responsible for the loss of host recognition by potatoes carrying the *Nx* resistance gene. Interestingly, an additional second-site mutation of Ile62 to Val resulted in a reversion back to avirulence. Based on the proposed coat protein structure of PVX, both of these residues may reside in close proximity to one another.[66] Thus, the virulence mutation, Glu78 to Pro, may interfere with the structure or the accessibility of a motif important for host recognition, whereas the second mutation at residue 62 may act to restore the structure of the recognition motif or its accessibility to the host component involved. Future studies comparing the biophysical properties of these two coat proteins should reveal how these mutations affect coat protein quaternary and tertiary structure.

Rx Derived Resistance

Rx resistance in potato provides effective resistance against most PVX strains, with the exception of PVX_{HB}.[67] *Rx* resistance usually results in no visible necrosis and no detectable virus accumulation in planta. The ability of PVX_{HB} to overcome

Rx resistance also maps to the coat protein ORF.[64] Kölem et al. found that a PVX coat protein frame shift mutation, resulting in the production of a truncated coat protein, compromised *Rx* mediated resistance, indicating that resistance is derived from the recognition of PVX coat protein.[68] Additionally, PVX activation of *Rx* resistance in potato protoplasts conferred levels of resistance against PVX_{HB} and cucumber mosaic virus (CMV). This generalized resistance is reminiscent of systemic acquired resistance (SAR) associated with the HR.[69] Additional information also indicates that PVX graft inoculations of Rx resistant stocks results in the formation of necrosis.[70] Thus, *Rx* resistance may represent the HR at a cellular level or the activation of a secondary plant mechanism involved in SAR.

Further analysis of the *RX* resistance-breaking strain PVX_{HB} demonstrated that a single coat protein amino acid substitution, Thr to Lys, at position 121 was responsible for overcoming *Rx* resistance.[71] This substitution impairs virus accumulation and spread. However, the addition of a second substitution identified in PVX_{HB}, Lys to Arg at position 127, acted to restore normal coat protein function involved in virus accumulation and spread. Thus, the ability of PVX_{HB} to overcome *RX* resistance is the result of two coat protein alterations, one affecting host recognition but adversely affecting coat protein function and a second alteration that apparently compensates for the first by restoring coat protein function. The Thr 121 to Lys substitution is also responsible for induction of the HR in *Gomphrena globosa,* indicating that a functionally homologous recognition mechanism exists between *G. globosa* and the *Rx* gene of potato.[72]

GENE VI OF CAULIMOVIRUSES

Cauliflower mosaic caulimoviruses (CaMV) are generally limited in host range to the cruciferae.[73] However, some strains can infect solanaceous species, causing either systemic disease or hypersensitive resistance.[74,75] For example, strains CM1841 and Cabb-B induce an HR in *Datura stramonium* and *Nicotiana bigelovii,* whereas strain D4 systemically infects both hosts. In order to map viral sequences responsible for HR elicitation, Schoelz et al. created a series of chimeric CaMV constructs between strains D4, CM1841, and Cabb-B.[76] A 496 base pair region comprising the first half of the CaMV gene VI ORF was found to be the determinant responsible for either the HR induced by strains CM1841 and Cabb-B or the systemic symptoms caused by strain D4. Between the strains there were 20 amino acid differences within this region that could have resulted in the altered phenotype. Thus, the CaMV gene VI protein may act as an elicitor in certain solanaceous hosts. Gene VI is also known to function as a trans-activator for the expression of other CaMV genes.[77,78] Therefore, alterations in gene VI may affect the expression of other CaMV gene products that act as determinants for HR induction. However, Takehashi et al. transformed tobacco with gene VI of the crucifer strain of CaMV.[79]. These transgenic plants displayed necrotic spots and expressed defenses-related proteins. These data provide further support for the gene VI protein acting as a specific elicitor.

LOCAL LESION DETERMINANTS IN TOMATO BUSHY STUNT TOMBUSVIRUS

The tomato bushy stunt tombusvirus (TBSV) genome encodes five genes. The two 3' distal genes, p22 and p19, encode products of 22 kDa and 19 kDa, respectively.[80] The p19 gene is nested within the p22 coding region. The functions of these genes have not been clearly established, although the p22 gene may be involved in cell-to-cell movement.[81] Scholthof *et al.,* have utilized a heterologous PVX expression vector[25] to systemically express genes of TBSV in different hosts[26]. These experiments have identified the p19 and p22 genes as determinants for the induction of necrotic local lesions in different tobacco species. p19 was found responsible for the induction of lesions in *N. clevelandii, N. benthamiana,* and *N. tabacum,* while p22 induced necrotic lesions in *N. glutinosa.* Thus, different *Nicotiana* species may target different viral genes for recognition.

LOCAL LESION DETERMINANTS IN BARLEY STRIPE MOSAIC HORDEIVIRUS

Different strains of barley stripe mosaic hordeivirus (BSMV) induce different local lesion phenotypes in *Chenopodium amaranticolor* Coste and Reyn. For example, the Type strain induces chlorotic lesions 14 days postinoculation while strain ND18 elicits necrotic lesions 3–4 days postinoculation.[82,83] It is not clear if these responses are due to a HR, since the ND18 induced lesions continue to expand and eventually collapse the entire inoculated leaf.[82] However, this system serves as an interesting example for demonstrating the complexities that can occur in virus-host interactions.

The tripartite nature of the BSMV genome, consisting of alpha, beta, and gamma RNAs, has allowed the use of pseudorecombinants to analyze genetic determinants involved in this phenotypic difference. Petty et al. tested all possible genomic combinations between Type and ND18 BSMV strains.[84] The ability to induce necrotic local lesions in *C. amaranticolor* was mapped to ND18 genomic RNA-gamma. RNA-gamma encodes for two protein products, gamma-a, a putative polymerase, and gamma-b, possibly involved in gene regulation. Analysis of genetic elements within RNA-gamma indicated that the leader ORF present in the Type strain is a determinant of the attenuated "chlorotic" local lesion phenotype. In vitro translation studies indicate that this leader decreases the efficiency of gamma-a protein translation,[85] suggesting that lower levels of gamma-a protein lead to the chlorotic local lesion phenotype induced by the Type strain.

It was also observed that chimeric RNA-gamma constructs containing the ND18 gamma-b protein ORF induced necrotic lesions in *C. amaranticolor* but at a timing similar to the chlorotic lesions induced by the parent Type strain.[84] Further analysis of chimeric mutants indicated the presence of several nucleotide differences in this region between ND18 and Type strains. Subsequent mutagenesis studies revealed a single point mutation leading to an amino acid substitution of Pro to Leu in the gamma-b protein was required for the formation of necrotic local lesions. The gamma-b protein has a putative virus gene regulatory function.[86] Thus, the identified amino acid substitution may indirectly affect the accumulation of other viral proteins.

The source of the alpha and beta genomic RNAs was also found to affect lesion phenotype.[84] For example, the necrotic lesion-forming ability of ND18 gamma RNA segregated cleanly in homologous backgrounds of RNA alpha and beta. However, various mutants of Type strain gamma RNA conferred different chlorotic and necrotic lesion phenotypes relative to the source of RNAs alpha and beta. Additionally, heterologous combinations of genomic alpha and beta RNAs also affected the lesion phenotype conferred by RNA gamma. Taken together, these findings demonstrate that multiple determinants ranging from gene regulation to genomic backgrounds can influence local lesion phenotypes.

VIROID-INDUCED NECROSIS

To date, identified activation of host defense responses appears to be restricted to viral protein products. However, plant viral nucleic acid, such as single and double stranded (ds) RNA, contain a high-degree of secondary structure important for viral processes including replication and virion assembly. Within mammalian systems, regulatory components like the P68 protein kinase are known to be activated by dsRNA,[87] triggering a series of reactions that effectively regulate protein synthesis.[88] Hiddinga et al. have identified a possible plant homologue to the mammalian P68 molecule that is phosphorylated by potato spindle tuber viroid (PSTV).[89] Diener et al. have since demonstrated that PSTV is capable of binding and activating the mammalian P68 protein in in vitro studies.[90] Of interest, P68 activation was ten-fold higher with a severe PSTV strain, causing systemic necrosis, then with a mild PSTV strain, causing no necrosis. The mild and severe strains differ by only two nucleotides. Viroids are the smallest known autonomously replicating disease agents, consisting of covalently closed circular RNAs (246 to 375 nucleotides) with a highly based paired secondary structure that encodes for no known protein.[91] Thus, they are completely dependent upon host encoded factors for replication. While the viroid induced necrotic response is not a true HR, these data clearly suggest that differential activation of host regulatory systems can occur via nucleic acid. Thus, the possibility exists that host responses, such as viral induced resistance, may also occur via nucleic acid recognition.

SUMMARY

From the available information it is clear that plants have evolved sophisticated pathogen surveillance mechanisms that recognize different viral components. For example, the tobamovirus genome encodes three main components: coat protein, movement protein, and replicase proteins. All of these components, in the proper host background, can serve as HR elicitors: Coat protein in *N'* gene tobacco and L^2 or L^3 gene peppers; movement protein in *Tm2* and $Tm2^2$ tomatoes; and replicase protein in *N* gene tobacco. This diversity for recognition is not unexpected, because viruses are entirely dependent upon their hosts for the cellular machinery needed to complete their life-cycles. Thus, at every life-cycle step precise interactions between virus and host components exist. In speculation, plant pathogen recognition systems may have evolved from these virus-host interactions.

We are just beginning to develop an understanding of how plant viruses elicit host resistance responses. Much of this information has been restricted to the identification of viral gene sequences or protein products that are responsible for HR elicitation. However, existing information clearly demonstrates the precise nature of elicitor recognition and HR induction. Many of the identified viral elicitors become active or inactive upon single amino acids substitutions. These changes primarily affect only elicitor recognition and not the viral function of the protein. The ability to overcome resistance may thus be dependent upon the ability of the virus to alter or mask its elicitor component without adversely affecting its viral function. Further work describing the structural and biophysical consequences of elicitor alterations will be critical in future attempts to define HR specificity and to determine its durability within different systems. Combining this understanding with information on the function of plant resistance genes will certainly result in novel approaches to utilize the HR on a wider scale for disease control.

ACKNOWLEDGMENTS

This work was supported in part by USDA grant 94–37303–0732.

REFERENCES

1. Allard, H. A. 1914. The mosaic disease of tobacco. *U.S. Dept. Agr. Bull.* 40.
2. Stakman, E. C. 1915. Relation between *Puccinia graminis* and plants highly resistant to its attack. J. Agric. Res. **4:**193–299.

3. Fritig, B., S. Kauffman, B. Dumas, P. Geoffroy, M. Kopp, and M. Legrand. 1987. Mechanisms of the hypersensitive reaction in plants. In *Plant Resistance to Viruses,* ed. B. D. Harrison, pp. 92–108. Wiley, Chichester.
4. Bol, J. F., H. J. M. Linthorst, and B. J. C. Cornelissen. 1990. Plant pathogenesis-related proteins induced by virus infection. *Annu. Rev. Phytopathol.* **28:**113–138.
5. Linthorst, H. J. M. 1991. Pathogenesis-related proteins of plants. *Critical Rev. Plant Sci.* **10:**123–150.
6. Flor, H. H. 1971. Current status of the gene-for-gene concept. *Annu. Rev. Phaytopathol.* **9:**275–296.
7. Keen, N. T. 1990. Gene-for-gene complementarity in plant-pathogen interactions. *Annu. Rev. Genet.* **24:**447–463.
8. Lamb, C. J., M. A. Lawton, M. Dron, and R. A. Dixon. 1989. Signals and transduction mechanisms for activation of plant defenses against microbial attack. *Cell* **56:**215–224.
9. Gabriel, D. W. and B. G. Rolfe. 1990. Working models of specific recognition in plant-microbe interactions. *Annu. Rev. Phytopathol.* **28:**365–391.
10. Keen, N. T. 1992. The molecular biology of disease resistance. *Plant Mol. Biol.* **19:**109–122.
11. Moffat, A. S. 1994. Mapping the sequence of disease resistance. *Science* **265:**1804–1805.
12. Bent, A. F., B. N. Kunkel, D. Dahlbeck, K. L. Brown, R. Schmit, J. Giraudat, J. Leung, and B. J. Saskawicz, 1994. RPS2 of *Arabidopsis thaliana* represents a new class of disease resistance genes. *Science* **265:**1856–1860.
13. Mindrinos, M., F. Katagiri, G. L. Yu, and F. M. Ausubel, 1994. The *A. thaliana* disease resistance gene RPS2 encodes a protein containing a nucleotide-binding site and leucine-rich repeats. *Cell* **78:**1089–1099.
14. Whitham, S., S. P. Dinesh-Kumar, D. Choi, R. Hehl, C. Corr, and B. Baker. 1994. The product of the tobacco mosaic virus resistance gene *N:* similarity to Toll and interleukin-1 receptor. *Cell* **78:**1101–1115.
15. Holmes, F. O. 1938. Inheritance of resistance to tobacco-mosaic disease in tobacco. *Phytopathology* **28:**553–561.
16. Holmes, F. O. 1946. A comparison of the experimental host ranges of tobacco-etch and tobacco mosaic viruses. *Phytopathology* **36:**643–659.
17. Fraser, R. S. S. 1988. Virus recognition and pathogenicity: Implications for resistance mechanisms and breeding. *Pestic. Sci.* **23:**267–275.
18. Fraser, R. S. S. 1990. The genetics of resistance to plant viruses. *Annu. Rev. Phytopathol.* **28:**179–200.
19. Provvidenti, R. and R. O. Hampton. 1992. Sources of resistance to viruses in the *Potyviridae. Arch. Virol.* **5:**189–211.
20. Rao, A. L. N. and R. I. B. Francki. 1982. Distribution of determinants for symptom production and host range on the three RNA components of cucumber mosaic virus. *J. Gen. Virol.* **61:**197–205.

21. Okuno, T., C. Hiruki, D. V. Rao, and G. C. Figueiredo. 1983. Genetic determinants distributed in two genomic RNAs of sweet clover necrotic mosaic, red clover necrotic mosaic and clover primary leaf necrosis viruses. *J. Gen. Virol.* **64:**1907–1914.

22. Gardner, R., A. Howarth, P. Hahn, M. Brown-Leudi, R. J. Shepherd, and J. Messing. 1981. The complete nucleotide sequence of an infectious clone of cauliflower mosaic virus by M13mp7 shotgun sequencing, *Nucleic Acids Res.* **9:**2871–2888.

23. Ahlquist, P. and M. Janda. 1984. cDNA cloning and in vitro transcription of the complete brome mosaic virus genome. *Mol. Cell. Biol.* **4:**2876–2882.

24. Dawson, W. O., D. L. Beck, D. A. Knorr, and G. L. Grantham. 1986. cDNA cloning of the complete genome of tobacco mosaic virus and production of infectious transcripts. *Proc. Natl. Acad. Sci. USA* **83:**1832–1836.

25. Chapman, S., T. Kavanagh, and D. Baulcombe. 1992. Potato virus X as a vector for gene expression in plants. *The Plant Journal* **2:**549–557.

26. Scholthof, H. B., K. B. Scholthof, and A. O. Jackson. 1994. Identification of tomato bushy stunt virus host-specific symptom determinants by expression of individual genes from a potato virus X vector. *The Plant Cell* **7:**1157–72.

27. Kunkel, L. O. 1932. Local lesions in aucube mosaic of tomato. *Phytopathology* **22:**16.

28. Funatsu, G. and H. Fraenkel-Conrat. 1964. Location of amino acid exchanges in chemically evoked mutants of tobacco mosaic virus. *Biochemistry* **3:**1356–1361.

29. Wittman, H. G. and B. Witman-Liebold. 1966. Protein chemical studies of two RNA viruses and their mutants. *Cold Spring Harbor Symp. Quant. Biol.* **31:**163–172.

30. Saito, T., T. Meshi, N. Takamatsu, and Y. Okada. 1987. Coat protein gene sequences of tobacco mosaic virus encodes host response determinant. *Proc. Natl. Acad. Sci. USA* **84:**6074–6077.

31. Knorr, D. A. and W. O. Dawson. 1988. A point mutation in the tobacco mosaic virus capsid protein gene induces hypersensitivity in *Nicotiana sylvestris. Proc. Natl. Acad. Sci. USA.* **85:**170–174.

32. Culver, J. N. and W. O. Dawson. 1989. Point mutations in the coat protein gene of tobacco mosaic virus induce hypersensitivity in *Nicotiana sylvestris. Mol. Plant-Microbe Interact.* **2:**209–213.

33. Culver, J. N. and W. O. Dawson. 1989. Tobacco mosaic virus coat protein: An elicitor of the hypersensitive reaction but not required for the development of systemic symptoms in *Nicotiana sylvestris. Virology* **173:**755–758.

34. Culver, J. N. and W. O. Dawson. 1991. Tobacco mosaic virus elicitor coat protein genes produce a hypersensitive phenotype in transgenic *Nicotiana sylvestris* plants. *Mol. Plant-Microbe Interact.* **4:**458–463.

35. Pfitzner, U. M. and A. J. P. Pfitzner. 1992. Expression of a viral avirulence gene in transgenic plants is sufficient to induce the hypersensitive defense reaction. *Mol. Plant-Microbe Interact.* **5:**318–321.

36. Bloomer, A. C., J. N. Champess., G. Bricogne, R. Staden, and A. Klug. 1978. Protein disk of tobacco mosaic virus at 2.8Å resolution showing the interactions within and between subunits. *Nature,* (London) **276:**362–368.

37. Namba, K., R. Pattaneyek, and G. Stubbs. 1989. Visualization of protein-nucleic acid interactions in a virus: refined structure of intact tobacco mosaic virus at 2.9Å resolution by X-ray fiber diffraction. *J. Mol. Biol.* **208:**307–325.

38. Durham, A. C. H., J. T. Finch, and A. Klug. 1971. States of aggregation of tobacco mosaic virus. *Nature,* (London) **229:**37–42.

39. Correia, J. J., S. Shire, D. A. Yphantis, and T. M. Schuster. 1985. Sedimentation equilibrium measurements of the intermediate-size tobacco mosaic virus protein polymers. *Biochemistry* **24:**3292–3297.

40. Raghavendra, K., J. A. Kelly, L. Khairallah, and T. M. Schuster. 1988. Structure and function of disk aggregates of the coat protein of tobacco mosaic virus. *Biochemistry* **27:**7583–7588.

41. Namba, K., D. L. D. Caspar, and G. Stubbs. 1989. Computer graphics representation of levels of organization in tobacco mosaic virus structure. *Science* **227:**773–776.

42. Culver, J. N., G. Stubbs, and W. O. Dawson. 1994. Structure-function relationship between tobacco mosaic virus coat protein and hypersensitivity in *Nicotiana sylvestris. J. Mol. Biol.* **242:**130–138.

43. Jockusch, H. 1966. Relations between temperature sensitivity , amino acid replacements, and quaternary structure of mutant proteins. *Biochem. Biophys. Res. Commun.* **24:**577–583.

44. Culver, J. N., W. O. Dawson, K. Plonk, and G. Stubbs. 1995. Site-directed mutagenesis confirms the involvement of carboxylate groups in the disassembly of tobacco mosaic virus. *Virology* **206:**724–730.

45. Saito, T., K. Yamanaka, Y. Watanabe, N. Takamatsu, T. Meshi, and Y. Okada. 1989. Mutational analysis of the coat protein gene of tobacco mosaic virus in relation to hypersensitive response in tobacco plants with the *N'* gene. *Virology* **173:**11–20.

46. Taraporewala, Z. F. and J. N. Culver. 1996. Identification of an elicitor active site within the three-dimensional structure of the tobacco mosaic tobamovirus coat protein. *The Plant Cell* **8:**169–178.

47. Altshcuh, D., A. M. Lesk, A. C. Bloomer, and A. Klug. 1987. Correlation of co-ordinated amino acid substitutions with function in viruses related to tobacco mosaic virus. *J. Mol. Biol.* **193:**693–707.

48. Tobias, I., R. S. S. Fraser, and A. Gerwitz. 1989. The gene-for-gene relationship between *Capsicum annuum L.* and tobacco mosaic virus - Effects on virus multiplication, ethylene synthesis and accumulation of pathogenesis-related proteins. *Physiol. Mol. Plant Pathol.* **35:**271–286.

49. Berzal-Herranz, A., A. De La Cruz, F. Tenllado, J. R. Diaz-Ruiz, L. Lopez, A. I. Sanz, C. Vaquero, M. T. Serra, and I. Garcia-Luque. 1994. The Capsicum L^3 gene-mediated resistance against the tobamoviruses is elicited by the coat protein. *Virology* **209:**498–505.

50. Dawson, W.O., P. Bubrick, and G. L. Grantham. 1988. Modifications of the tobacco mosaic virus coat protein gene affect replication, movement and symptomatology. *Phytopathology* **78:**783–789.

51. Tobias, I., A. T. B. Rast, and D. Z. Maat. 1982. Tobamoviruses of pepper, eggplant, and tobacco: Comparative host reactions and serological relationships. *Neth. J. Plant Pathol.* **88:**257–268.

52. Padgett, H. S. and R. N. Beachy. 1993. Analysis of a tobacco mosaic virus strain capable of overcoming N gene-mediated resistance. *The Plant Cell* **5:**577–586.

53. Deom, C. M., S. Wolf, C. A. Holt, W. J. Lucas, and R. N. Beachy. 1991. Altered function of the tobacco mosaic virus movement protein in a hypersensitive host. *Virology* **180**:251–256.

54. Pelham, J. 1972. Strain-genotype interaction of tobacco mosaic virus in tomato. *Ann. Appl. Biol.* **71:**219–228.

55. Hall, T. J. 1980. Resistance at the Tm-2 locus in the tomato to tomato mosaic virus. *Euphytica* **29:**189–197.

56. Motoyoshi, F. and N. Oshima. 1975. Infections with tobacco mosaic virus of leaf mesophyll protoplasts from susceptible and resistant lines of tomato. *J. Gen. Virol.* **29:**81–91.

57. Nishiguchi, M. and F. Motoyoshi. 1987. Resistance mechanisms of tobacco mosaic virus strains in tomato and tobacco. In *Plant Resistance to Viruses,* eds. D. Evered and S. Harnett, pp. 38–46, Wiley, Chichester.

58. Meshi, T., F. Motoyoshi, T. Maeda, S. Yoshiwoka, H. Watanabe, and Y. Okada. 1989. Mutations in the tobacco mosaic virus 30kd protein gene overcome TM-2 resistance in tomato. *The Plant Cell* **1:**515–522.

59. Weber, H., S. Schultze, and A. J. P. Pfitzner. 1993. Two amino acid substitutions in the tomato mosaic virus 30–kilodalton movement protein counter the ability to overcome the Tm-2^2 resistance gene in tomato. *J. Virology* **67:**6432–6438.

60. Gafny, R., M. Lapidot, A. Berna, C. A. Holt, C. M. Deom, and R. N. Beachy. 1992. Effects of terminal deletion mutations on function of the movement protein of tobacco mosaic virus. *Virology* **187:**499–507.

61. Weber, H., S. Schultze, and A. J. P. Pfitzner. 1994. The Tm-2^2 resistance gene in tomato requires the recognition of a specific domain of the tomato mosaic virus 30–kD movement protein. (Abstract) 4th International Congress of Plant Molecular Biology, The International Society for Plant Molecular Biology.

62. Cockerham, G. 1955. Strains of potato virus X. *Proceedings of the Conference on Potato Virus Diseases,* Lisse-Wageningen, 1954.

63. Adams, S. E., R. A. C. Jones, and R. H. A. Coutts. 1986. Effect of temperature on potato virus X infection in potato cultivars carrying different combinations of hypersensitivity genes. *Plant Pathol.* **35:**517–526.

64. Kavanagh, T., M. Goulden, S. Santa Cruz, S. Chapman, I. Barker, and D. C. Baulcombe. 1992. Molecular analysis of a resistance-breaking strain of potato virus X. *Virology* **189:**609–617.

65. Santa Cruz, S. and D. C. Baulcombe. 1993. Molecular analysis of potato virus X isolates in relation to the potato hypersensitivity gene *Nx. Mol. Plant-Microbe Interact.* **6:**707–714.

66. Baratova, L. A., N. I. Grebenshchikov, E. N. Dobrov, A. V. Gedrovich, I. A. Kasharin, A. V. Shishkov, A. V. Efimov, L. Järvekulg, Y. L. Radavsky, and M. Saarma. 1992. The organization of potato virus X coat proteins in virus particles studied by tritium planigraphy and model building. *Virology* **188:**175–180.

67. Moreira, A., R. A. C. Jones, and C. E. Fribourg. 1980. Properties of a resistance-breaking strain of potato virus X. *Ann. Appl. Biol.* **95:**93–103.

68. Köhm, B. A., M. G. Goulden, J. E. Gilbert, T. A. Kavanagh, and D. C. Baulcombe. 1993. A potato virus X resistance gene mediates an induced, nonspecific resistance in protoplasts. *The Plant Cell* **5:**913–920.

69. van Loon, L. C. and J. Dijkstra. 1976. Virus-specific expression of systemic acquired resistance in tobacco mosaic virus- and tobacco necrosis virus-infected "Samsun *NN*" and "Samsun" tobacco. *Neth. J. Plant Pathol.* **82:**231–237.

70. Benson, A. P. and W. J. Hooker. 1960. Isolation of virus X from "immune" varieties of potato, *Solanum tuberosum. Phytopathology* **50:**231–234.

71. Goulden, M. G., B. A. Köhm, S. Santa Cruz, T. A. Kavanagh, and D. C. Baulcombe. 1993. A feature of the coat protein of potato virus X affects both induced virus resistance in potato and viral fitness. *Virology* **197:**293–302.

72. Goulden, M. G. and D. C. Baulcombe. 1993. Functionally homologous host components recognize potato virus X in *Gomphrena globosa* and potato. *The Plant Cell* **5:**921–930.

73. Walker, J. C., F. J. LeBeau, and G. S. Pound. 1945. Viruses associated with cabbage mosaic. *J. Agric. Res.* **70:**379–404.

74. Lung, M. C. Y. and T. P. Pirone. 1972. Datura stramonium, a local lesion host for certain isolates of cauliflower mosaic virus. *Phytopathology* **62:**1473–1474.

75. Schoelz, J. E., R. J. Shepherd, and R. D. Richins. 1986. Properties of an unusual strain of cauliflower mosaic virus. *Phytopathology* **76:**451–454.

76. Schoelz, J., R. J. Shepherd, and S. Daubert. 1986. Region VI of cauliflower mosaic virus encodes a host range determinant. *Mol.Cell. Biol.* **6:**2632–2637.

77. Booneville, J. M., H. Sanfacon, J. Futterer, and T. Hohn. 1989. Post-transcriptional trans-activation in cauliflower mosaic virus. *Cell* **59:**1135–1143.

78. Gowda, S., F. C. Wu, H. B. Scholtof, and R. J. Shepherd. 1989. Gene VI of figwort mosaic virus (caulimovirus group) functions in post-transcriptional expression of genes on the full-length RNA transcript. *Proc. Nat. Acad. Sci. USA* **86:**9203–9207.

79. Takahashi, H., K. Shimamoto, and Y. Ehara. 1989. Cauliflower mosaic virus gene VI causes growth suppression, development of necrotic spots and expression of defense-related genes in transgenic tobacco plants. *Mol. Gen. Genet.* **216:**188–194.

80. Hearne, P. Q., D. A. Knorr, B. I. Hillman, and T. J. Morris. 1990. The complete genome structure and synthesis of infectious RNA from clones of tomato bushy stunt virus. *Virology* **177:**141–151.

81. Rochon, D. M. and J. C. Johnston. 1991. Infectious transcripts from cloned cucumber necrosis virus cDNA: Evidence for a bifunctional subgenomic mRNA. *Virology* **181:**656–665.

82. Edwards, M. C. and R. G. Timian. 1986. Genetic analysis of barley stripe mosaic virus. *Phytopathology* **76:**360–365.
83. Petty, I. T. D., B. G. Hunter, N. Wei, and A. O. Jackson. 1989. Infectious barley stripe mosaic virus RNA transcribed *in vitro* from full-length genomic cDNA clones. *Virology* **171:**242–249.
84. Petty, I. T. D., R. G. K. Donald, and A. O. Jackson. 1994. Multiple genetic determinants of barley stripe mosaic virus influence lesion phenotype on *Chenopodium amaranticolor. Virology* **198:**218–226.
85. Petty, I. T. D., M. C. Edwards, and A. O. Jackson, 1990. Systemic movement of an RNA plant virus determined by a point substitution in a 5' leader sequence. *Proc. Natl. Acad. Sci. USA* **87:**8894–8897.
86. Petty, I. T. D., R. French, R. W. Jones, and A. O. Jackson. 1990. Identification of barley stripe mosaic virus genes involved in viral RNA replication and systemic movement. *EMBO J.* **9:**3453–3457.
87. Meurs, E., K. Chong, J. Galabru, N. S. B. Thomas, I. M. Kerr, B. R. G. Williams, and A. G. Hovanessian. 1990. Molecular cloning and characterization of the human double-stranded RNA-activated protein kinase induced by interferon. *Cell* **62:**379–390.
88. Hershey, J. W. B. 1991. Translational control in mammalian cells. *Annu. Rev. Biochem.* **60:**717–755.
89. Hiddinga, H. J., C. J. Grum, J. Hu, and D. A. Roth, 1988. Viroid-induced phosphorylation of a host protein related to a dsRNA-dependent protein kinase. *Science* **241:**451–453.
90. Diener, T. O., R. W. Hammond, T. Black, and M. G. Katze. 1993. Mechanism of viroid pathogenesis: Differential activation of the interferon-induced, double-stranded RNA-activated, M_r68,000 protein kinase by viroid strains of varying pathogenicity. *Biochimie* **75:**533–538.
91. Diener, T. O. 1991. Subviral pathogens of plants: viroids and viroidlike satellite RNAs. *FASAB J.* **5:**2808–2813.

Index

MIX
Papier aus verantwortungsvollen Quellen
Paper from responsible sources
FSC® C105338

If you have any concerns about our products,
you can contact us on
ProductSafety@springernature.com

In case Publisher is established outside the EU,
the EU authorized representative is:
Springer Nature Customer Service Center GmbH
Europaplatz 3, 69115 Heidelberg, Germany

Printed by Libri Plureos GmbH
in Hamburg, Germany